MW01715046

Conceived and produced by
Lionheart Books
10 Chelmsford Square
London NW10 3AR

Editor: Lionel Bender
Designer: Ben White
Assistant Editor: Madeleine Samuel

Copyright © Lionheart Books 1988

© 1989 – this adaptation Hampstead Press, New York/Toronto

Adapted and published in the United States in 1989 by Hampstead Press, 387 Park Ave. South, New York NY 10016

All rights reserved

Printed and bound in Belgium by Proost International Book Production

Bramwell, Martyn.
Oceanography/Martyn Bramwell.
p. cm. — (Science frontiers)
Includes index.
Summary: Discusses technological developments affecting the field of oceanography and the use of marine resources.
ISBN 0-531-19510-4
1. Oceanography — Juvenile literature.
[1. Oceanography. 2. Technology.]
I. Title. II. Series.
GC21.5.B72 1989
551.46–dc 19 88-25907
CIP
AC

Illustrations by Hayward Art Group, except pages 16–17, 24–25, 32–33 and 40–41, by James G. Robins.

Make up: Radius

Picture credits
Pages as numbered. T=top, C=center, B=bottom, L=left, R=right. [Planet Earth Pictures (PEP). Science Photo Library (SPL)].
5: ZEFA, 6: Canada House, 8: PEP/John Menzies. 9: PEP/Al Giddings, 10: Institute of Oceanographic Studies, Wormley, England (IOS), 11T: Ocean Drilling Program/Texas A&M Univ., 11B: IOS, 12T: Woods Hole Oceanographic Institution, 12B: Japan Marine Science and Technology Center, 13T: PEP/James King, 13B: PEP/Ian Vaughan, 14: PEP/Christian Petron, 15T: SPL Stan Wayman, 15C: Woods Hole Oceanographic Institution, 15BL,BR: National Maritime Museum, Greenwich, 18T: SPL/Dr Legeckis, 18B: PEP/J Duncan, 19T: Rauma Repola, Finland, 19B: Paul Rodhouse/ British Antarctic Survey, 20T: Scripps Institution of Oceanography, San Diego, 20B: Canadian Dept. of Fisheries and Oceans, 21T: Baltimore Gas and Electric Co., 21B: Canadian Dept. of Fisheries and Oceans, 22: PEP/Robert Jureit, 23T: PEP/Ken Lucas, 23B: Torry Research Station, Crown Copyright, 26: A Shell Photograph, 27: Kvaenar Brug A/S, 28C: PEP/Ivor Edmonds, 28B: Ocean Drilling Program/Texas A&M Univ., 29T: PEP/Robert Hessler, 29B: SPL/Peter Ryan, Scripps Instit., 30T: PEP/David Maitland, 30B: PEP/Ken Lucas, 31T: Canadian Dept., of Fisheries and Oceans, 31B: PEP/Robert Hessler, 34T: PEP/Rob Beighton, 34B: BP Chemicals Ltd (RIGIDOIL is a registered trade mark of The BP Co. Ltd.), 35T: Courtesy of EASAMS Ltd., 35B: Prof. T.A. Norton, Port Erin, Isle of Man, 36C: Offshore Engineer, 36B: Hydraulics Research, 37T: Aer°Camera-BartHofmeester bv Rotterdam, 37C: Offshore Engineer, 38C: Peter Prince, British Antarctic Survey, 38B: Greenpeace Films Ltd., 39T: ZEFA, 39B: PEP/Mike Coltman, 42–43T: Rauma Repola, Finland, 42B: Canadian Dept. of Fisheries and Oceans, 43T: Courtesy of The British Petroleum Company plc. 43B: PEP/Christian Petron, Cover: PEP/Peter Scoones.

Cover photo: Scuba diver beneath offshore rig.

Photo opposite: Diver photographing corals in Red Sea.

SCIENCE FRONTIERS

Oceanography

MARTYN BRAMWELL

Hampstead Press New York 1989

KENMORE EAST H S LIBRARY

ABOUT THIS BOOK

Science Frontiers reviews the current state of scientific research and development in the major areas of technology. Its aim is to show what scientists, inventors and designers are trying to achieve, and to explain why their work is so important. Why, for example, are medical scientists so excited about Caribbean sponges and sea squirts? Or about the blood of the horseshoe crab? What can a microscopic shrimp tell us about the North Sea?

This book is divided into four sections—Exploring the Deep, Farming the Waters, The Oceans' Resources and Managing the Oceans. Each describes the major trends in that particular field of oceanographic science or technology, and then looks in more detail at selected examples of research and development. The stories are illustrated throughout with photographs and diagrams. The final part of each section projects current trends, and shows how familiar settings such as fishing ports and coastguard stations might look in the year 2001.

At the end of the book, What Next? looks a little beyond 2001. The glossary provides definitions of technical words used in the text.

△ Drilling for oil from an artificial island in the Beaufort Sea, Canada.

CONTENTS

INTRODUCTION 8
STUDYING THE SEABED 10
SHIPS, SUBS AND DIVERS 12
ARCHAEOLOGY UNDER WATER 14
WORKING BENEATH THE WAVES 16
HUNTERS AND CATCHERS 18
SAFEGUARDING THE STOCKS 20
SEA FOODS OF THE FUTURE 22
THE OCEAN HARVEST 24
ENERGY FROM THE SEA 26
WATER AND MINERALS 28
MARINE BIOTECHNOLOGY 30
SHARING THE OCEAN RESOURCE 32
CLEANING UP OUR ACT 34
CONTROLLING LAND AND SEA 36
PROTECTING THE MARINE WORLD 38
A NEW DEAL FOR THE OCEANS 40
NEW FRONTIERS 42
GLOSSARY 44
INDEX 45

INTRODUCTION

People have been fascinated by the oceans for thousands of years. Seamanship dates back 40,000 years or more. Aristotle was writing about the salinity of the sea over 2,300 years ago. In 1665, the Royal Society in London issued instructions to seafarers asking them to collect information about tides, currents and water depths wherever they went. And in 1872, HMS *Challenger* set sail on her four-year around-the-world voyage of discovery. It was the world's first scientific oceanographic expedition.

Despite all this, the deep oceans remained largely unknown until well into the 20th century. Collecting samples really was a case of groping in the dark. But everything changed after World War II. Radio, radar and sonar had appeared on the scene. More and more scientists and technicians joined the oceanographic organizations. In fewer than 50 years oceanography and its related sciences have been revolutionized. Satellites, micro-electronics, heavy engineering and biotechnology are now all part of the story. We are on the brink of major advances in food technology, energy supply, bio-medicine and genetic engineering. We are also able to manage the ocean environment and conserve marine life—if we chose.

▽ A submersible is lowered into the sea from its support ship. Vessels like this are used to service, inspect and repair undersea oil and gas pipelines.

▷ The Johnson-Sea-Link submersible is used for both scientific and commercial work. It can handle tools and small items of equipment with precision.

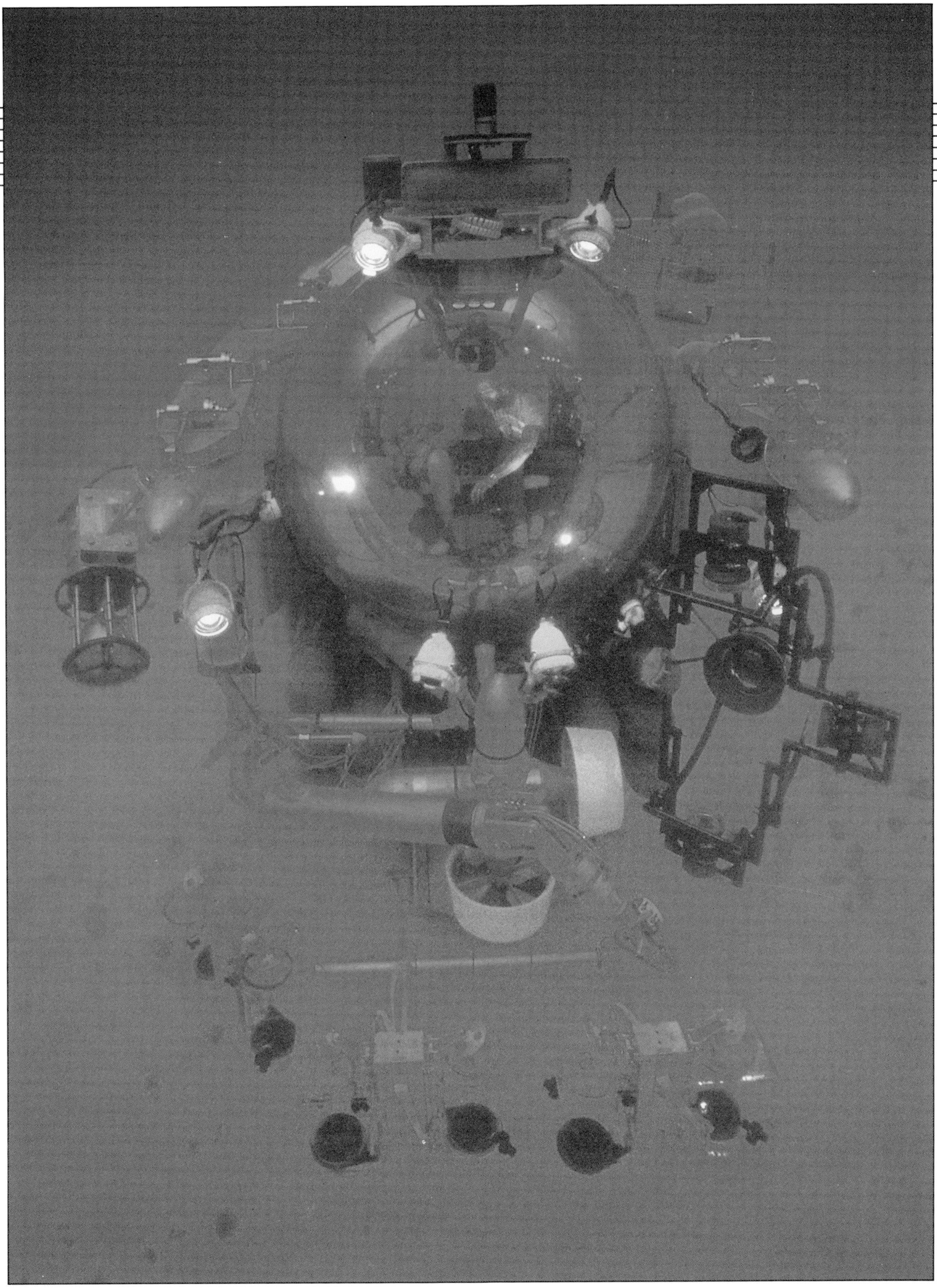

STUDYING THE SEA BED

In the past, the only way of studying the oceans was by lowering weighted measuring lines, nets, grappling hooks and bottles into the water. Now it is possible to photograph the deep ocean bed, or create pictures of its mountains, valleys and plains, using sound waves.

Instrument packages sent down to the ocean bed may sit there for months, measuring temperatures, currents and submarine earth tremors. Some transmit their information to the surface. Others store it on tape until the package is ordered and return to the surface by an acoustic signal from a ship.

▼ GLORIA SOUNDS OUT THE DEEP

GLORIA (Geological Long Range Inclined Asdic) is the world's most advanced echo sounder system. It was developed in Britain at the Institute of Oceanographic Sciences (IOS), and has been used all over the world for mapping the ocean bed and for studying the geology of ocean basins.

GLORIA records are extremely detailed and have provided scientists with dramatic pictures of mid-ocean mountain ranges, submarine volcanoes and faults, and huge mud waves created by sea bed currents. GLORIA works by scanning the ocean bed with pulses of sound that fan out up to 30 km (20 mi) at either side. Receivers pick up echoes that bounce back from the ocean bed, and these acoustic signals are converted into electrical signals and then into pictures that are built up line by line as the ship moves forward.

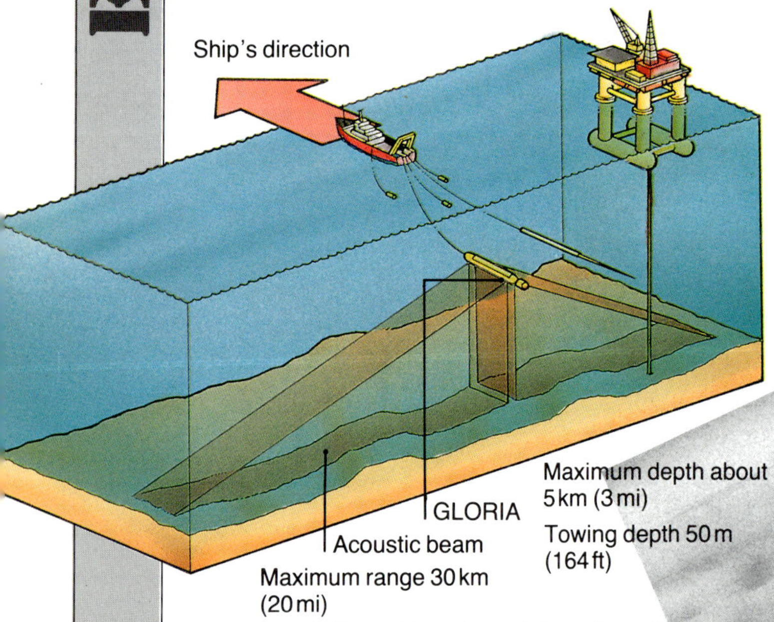

▲ EYE IN THE SKY

SEASAT was an experimental satellite launched in 1978 by NASA. It was in operation for only three months, yet produced so much information that scientists are still analyzing it. The biggest surprise came from SEASAT's radar altimeter, which measure the satellite's height to an accuracy of 10 cm (4 in). By analyzing this height information, scientists can study the shape of the Earth, and of the ocean floor. The European satellite ERS-1, due to be launched in 1989, will take this research even further.

△ GLORIA is nearly 8 m (26 ft) long and weighs 2 tons. It is carried in a special launch-and-recovery cradle mounted on the ship's stern. The equipment can be used at speeds up to 18 kph (11 mph) and this allows huge areas to be covered. In one cruise 650,000 km (250,000 sq mi) of ocean bed were mapped in detail in just 96 days.

▶ DEEP-SEA DRILLING

In 1984 the oil exploration ship *SEDCO/BP 471* was completely refitted and transformed into a floating scientific research station. The ship was renamed *JOIDES Resolution*, and its main mission is to lead a 10-year program of deep-sea drilling to discover more about the rock types and history of the ocean floor. The program is an international effort involving research institutions of 18 countries. JOIDES stands for Joint Oceanographic Institutions for Deep Earth Sampling.

Resolution is 143 m (469 ft) long and 21 m (69 ft) wide, and its drilling derrick towers high above the water line. Computer-controlled automatic positioning systems enable the ship to hold position accurately over a pin-prick in the sea bed, even in very deep water. The drilling system can suspend up to 9,000 m (5½ mi) of drill pipe through a hole in the ship's hull. This means that in water 4,000 m (2½ mi) deep, rock samples can be obtained from 5,000 m (3 mi) beneath the sea bed.

▼ POP-UP TECHNOLOGY

Pop-up technology is a new technique that can be used with many different research instruments. An instrument package complete with power supply, timing mechanism, recorders and other equipment, can be placed on the ocean bed and then left for weeks while the ship does other work. Readings from the instruments can be transmitted to the ship, or they can be recorded and stored on magnetic tape or wire. When the ship returns, an acoustic (sound) signal is beamed down to the instrument pack. Automatic catches release the instruments from the heavy iron ballast weight, and foam-filled buoyancy floats carry them to the surface.

Among the instruments used in this way is the Bathysnap camera. The device is set to take flash photographs every few minutes or hours over a period of time.

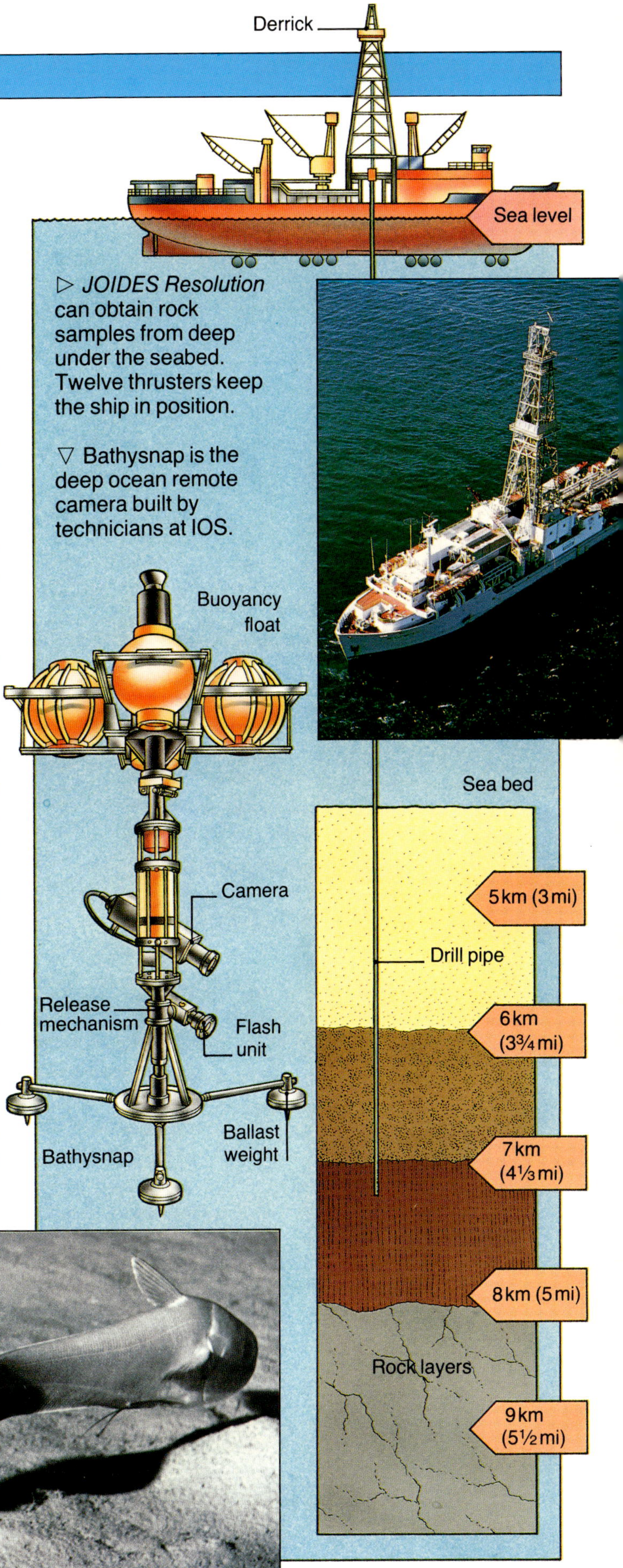

▷ *JOIDES Resolution* can obtain rock samples from deep under the seabed. Twelve thrusters keep the ship in position.

▽ Bathysnap is the deep ocean remote camera built by technicians at IOS.

▷ Bathysnap has produced pictures of sea urchins and sea cucumbers feeding on tiny food particles in the sea bed mud. It has photographed unknown sea bed worms and starfish. And it has captured many superb pictures of deep ocean fish, like this rat-tailed fish, taken at a depth of nearly a mile in the Atlantic.

SHIPS, SUBS AND DIVERS

Apart from their obvious use as military vessels, submarines (subs) have two very important uses today. They provide scientists with a safe means of studying the geology and biology of the deep ocean basins, and they are an essential part of the technology of the oil and gas industries.

The design of new submersibles (the word used for small non-military submarines) brings together many branches of science and technology. A great deal of medical research is needed to understand how a diver's body is affected by pressure and by breathing different mixtures of gases. Many different and unusual materials are used in the construction of the vessels themselves. New kinds of communications equipment have been developed for use underwater, and completely new motors, driven by pumped sea water, are now being designed. These will power the tools used by divers and remote controlled submersibles for cleaning and repairing pipelines.

▼ SHIP RESEARCH IN JAPAN

Japanese shipbuilders and marine technologies are developing many new designs for both surface and underwater research vessels. The Japanese Ship Research Institute is working closely with shipbuilders Mitsubishi and Mitsui on the development of SWATH vessels like the *Kaiyo*. These twin-hulled ships are extremely stable. They will roll from side to side 5° for example, in conditions that would make a single-hulled ship roll 30°. The ship has the most advanced positioning, navigation and noise-control facilities available.

JAMSTEC, the Japan Marine Science and Technology Center, is currently developing a new research submersible with a planned operating depth of 6,500 m (4 mi).

▽ At 61.5 m (200 ft) *Kaiyo* is the world's largest SWATH vessel. The name means Small Waterplane Area Twin Hull. It has many unusual features.

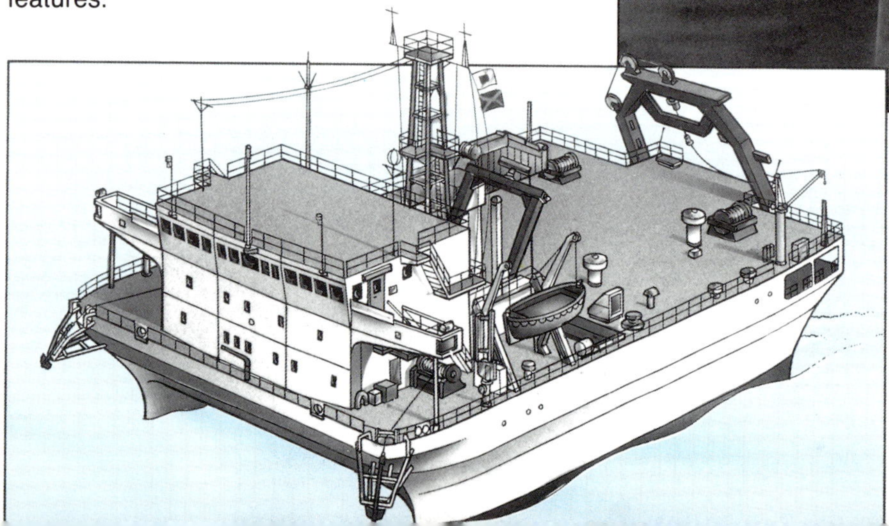

△ *Atlantis II* with the research submersible *Alvin* about to be lowered into the water from her stern.

▲ NEW SHIPS FOR OLD

As many of the world's older research ships reach retirement age, they are being replaced by vessels that look very different. Gone are the cramped decks and cluttered superstructures of ships such as Britain's RRS *Shackleton* and the United States' *Atlantis II*, support ship for the submersible *Alvin*. The new ships are designed by marine engineers and research scientists working together. They have open deck space with fixing points that will take a wide variety of equipment according to the purpose of each cruise. Below decks there are laboratories, wet and dry storage areas, and technical and scientific workshops. There are also custom built computer rooms.

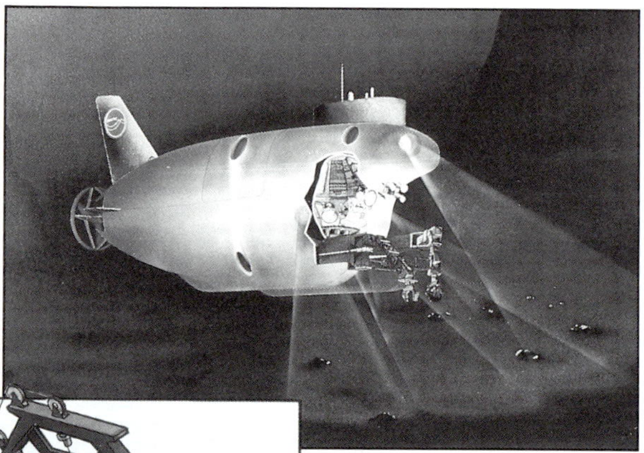

△ The new JAMSTEC submersible will help scientists working on the hot water vents of the ocean floor and on ways of predicting earthquakes.

ONE-PERSON SUBMERSIBLES

A one-person submersible, or atmospheric diving system, is like a cross between a submarine and a diving suit. It is a heavily armored human sized capsule, lowered into the water on a high strength steel cable. This cable also supports the power lines for the suit's thruster motors, and the various cables for video cameras, tools and telephone communications. Inside, the diver stands upright, controlling the thrusters by means of foot-operated switches. The diver places his or her arms inside the suit's jointed arms in order to operate the mechanical claws.

With this kind of system, the diver is automatically kept at normal atmospheric pressure throughout the dive so there is no need for compression or decompression. If the power fails, the suit is automatically switched over to emergency battery power that keeps the life support systems working properly.

△ Modern one-person subs like the WASP shown here developed from armored diving suits of the 1970s. These had jointed legs and arms, but no thrusters. WASP weighs 400 kg (900 lb) and can be used down to 700 m (2,300 ft).

MINI-SUBS AND ROVs

The British Oceanics LR2 submersible is typical of the multi-purpose commercial mini-subs that were developed for use in the North Sea oil and gas fields. The vehicles were used to survey pipeline routes and to carry out repair and maintenance work on the underwater steelwork that supports drilling rigs.

One very important job done by mini-subs is to remove the thick layers of algae, shells and chalky worm tubes that build up like concrete all over the steel. This work is now done mainly by diverless mini-subs called ROVs – Remote Operated Vehicles. They have no crew on board. Instead they are controlled through a command cable linking them to the support ship above. Television cameras mounted on the vehicle enable the pilot, far above, to see exactly what he or she is doing. Two of the latest cleaning machines are Scimitar and MOC-1. They were developed for use in Australia's warm waters where marine growth is a big problem.

◁ Since oil exploration and drilling work in the North Sea slowed down, mini-subs have been short of work. But some may now have a new career. There are plans to use mini-subs like the LR2 to carry tourists on sight-seeing trips around the spectacular coral reefs of the Caribbean.

ARCHAEOLOGY UNDERWATER

Until recently, archaeologists had to work almost entirely on land.

All this changed in the 1960s when Scuba (Self-Contained Underwater Breathing Apparatus) was developed. Marine archaeologists now use an enormous range of tools and equipment. Wrecks in shallows are excavated with suction pumps. Computers are used to record every piece of timber removed, and to plot each one's exact position. Side-scan sonar (a small version of GLORIA) is now used to pinpoint wreck sites. The *Defense*, a famous ship of the American Revolution, was found by this method, even though it projected only 30 cm (12 in) above the ocean bed.

But finding a wreck is only part of the story. Archaeologists are now helped by X-ray techniques that can reveal the coins, spoons and belt buckles stuck together in hard, rock-like lumps called concretions. And new techniques of wood preservation make it possible for millions of people to see ships like the *Mary Rose*, put back together after nearly 450 years under water.

△ Archaeologists record the position of a ship's timber using a reference grid of white strings.

▶ ALVIN AND JASON Jr

The Deep Submergence Research Vehicle (DSRV) *Alvin* is owned by the U.S. Navy and operated by the Woods Hole Oceanographic Institution. Since 1964, the vessel has made over 1,400 dives to depths of 4,000 m (2½ mi). *Alvin* has remote-controlled manipulator arms, cameras and lights, data collecting and recording systems, and instruments for measuring water temperatures, currents and the vehicle's depth. The pilot and two scientists work inside a titanium alloy pressure sphere.

▷ *Alvin* and *Jason Jr* are maneuvred through the ocean depths by means of sonar signals from beacons on the sea bed. By analyzing the sonar data, scientists on the support ship can monitor the depths and relative positions of the submersibles.

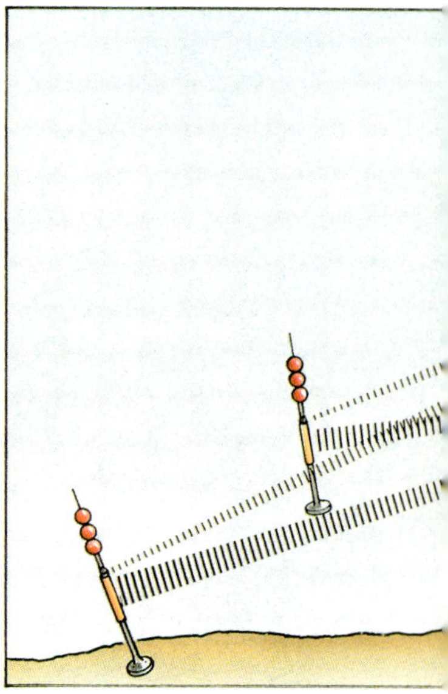

◀ HISTORY ON THE SEA BED

Every ship that ever went down in a storm, or was sunk in battle, took with it a piece of history. Like a time capsule, it carried down to the sea bed the weapons and navigation instruments, tools and clothing of the officers and crew. Also lost were the ships' cargos – in some cases tea, coffee or spices, and in others gems.

Technology has now brought these historical treasures within reach. Marine archaeologists mark out the wreck site with a grid of tapes or scaffold poles so that everything can be recorded. The site is photographed, often with 3–D cameras, then the excavation begins. Sand and mud is removed with suction hoses and dumped clear of the wreck. Air-filled bags are used to life heavy timbers or cannon to the surface. Special cutting and drilling tools are used to dismantle large sections of timberwork, finally, very large concretions may be broken up by expert use of small explosive charges without harming any of the objects they may contain.

Alvin now has a partner, *Jason Jr*. This robot is about the size of a portable TV set and is attached to *Alvin* by an 85 m (280 ft) command cable. The tiny robot carries lights and both video and still cameras. It has four thruster motors and is steered by the *Alvin* crew by remote control. In July 1986, *Alvin* and *Jason Jr* made world headlines by providing pictures of the liner *Titanic* lying at 3,800 m (12,500 ft) on the bed of the Atlantic. A new system consists of a large vehicle (Argo) flying 40 m (130 ft) above the sea bed with sonar and TV cameras, and a smaller tethered robot (*Jason*) that is sent out for close-up work.

△ *Alvin* is 7.6 m (22 ft) long and 2.8 m (8 ft) wide. Power is provided by a large propeller and two side-mounted thrusters.

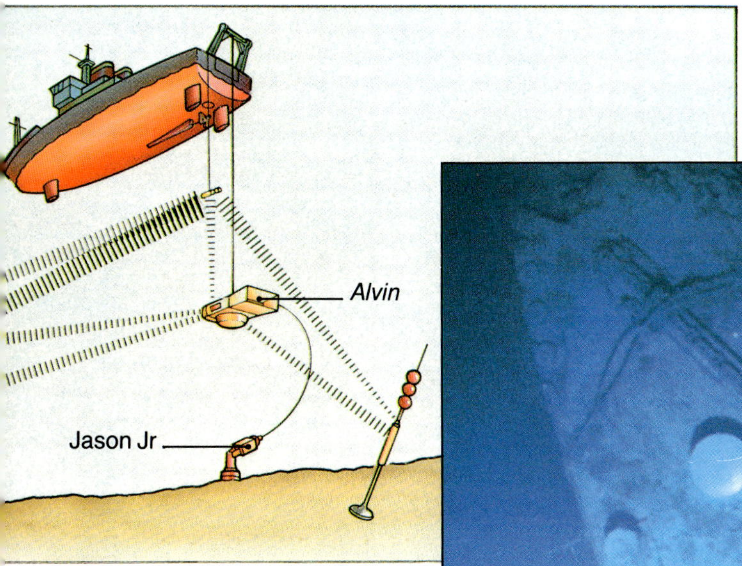

◁ A sad reminder of a great liner. This *Jason* photograph shows the bow section of the *Titanic* and her anchor chains.

PRESERVING THE FINDS

Wood that has been waterlogged for hundreds of years splits, shrinks and warps then it dries out. Without special treatment the timbers of a ship become twisted out of shape and ruined when exposed to the air. To preserve the wood, it is strengthened by replacing some of the water with a substance called polyethylene glycol (PEG).

▷ The effect of PEG treatment on waterlogged wood. In wood allowed to dry without treatment, the cells have been badly damaged (left-hand photo). In treated wood, the cells are supported by the PEG and have dried out with no splitting or twisting.

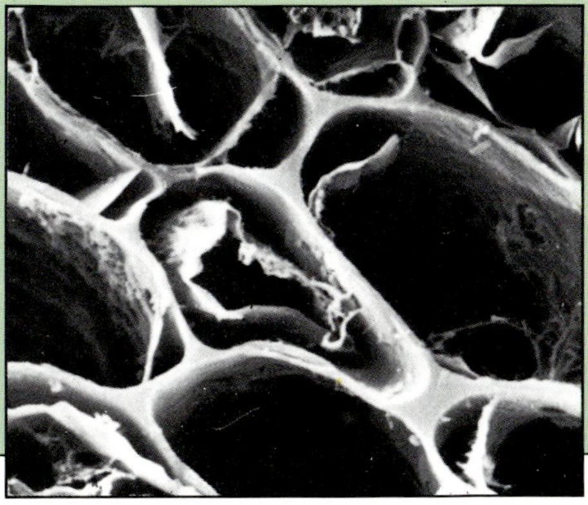

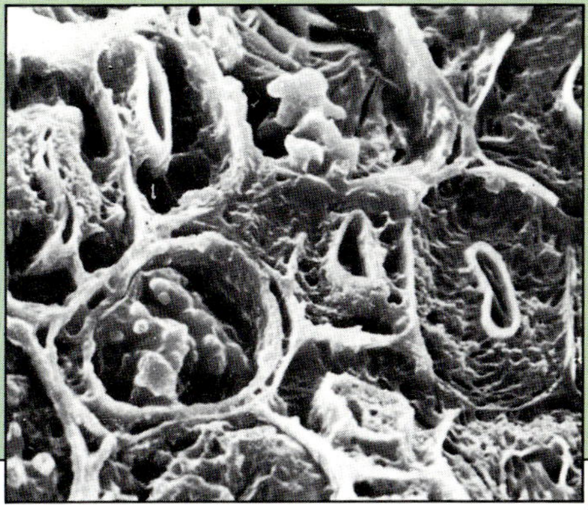

WORKING BENEATH THE WAVES

One of the most exciting developments in underwater technology is the Remote Operated Vehicle (ROV). There will always be a need for divers and manned submersibles, but for deep dives and dangerous underwater work the new ROVs are quicker, cheaper and safer. They can dive much deeper, stay down longer, and do almost anything a diver can do – with none of the risk. Small ROVs are now used to check ship's propellers, look for lost equipment, and inspect fish farms, piers and harbors for damage. Larger versions equipped with manipulators and special tools are taking over many underwater maintenance and salvage jobs, and are providing ocean scientists with new ways of exploring the deep.

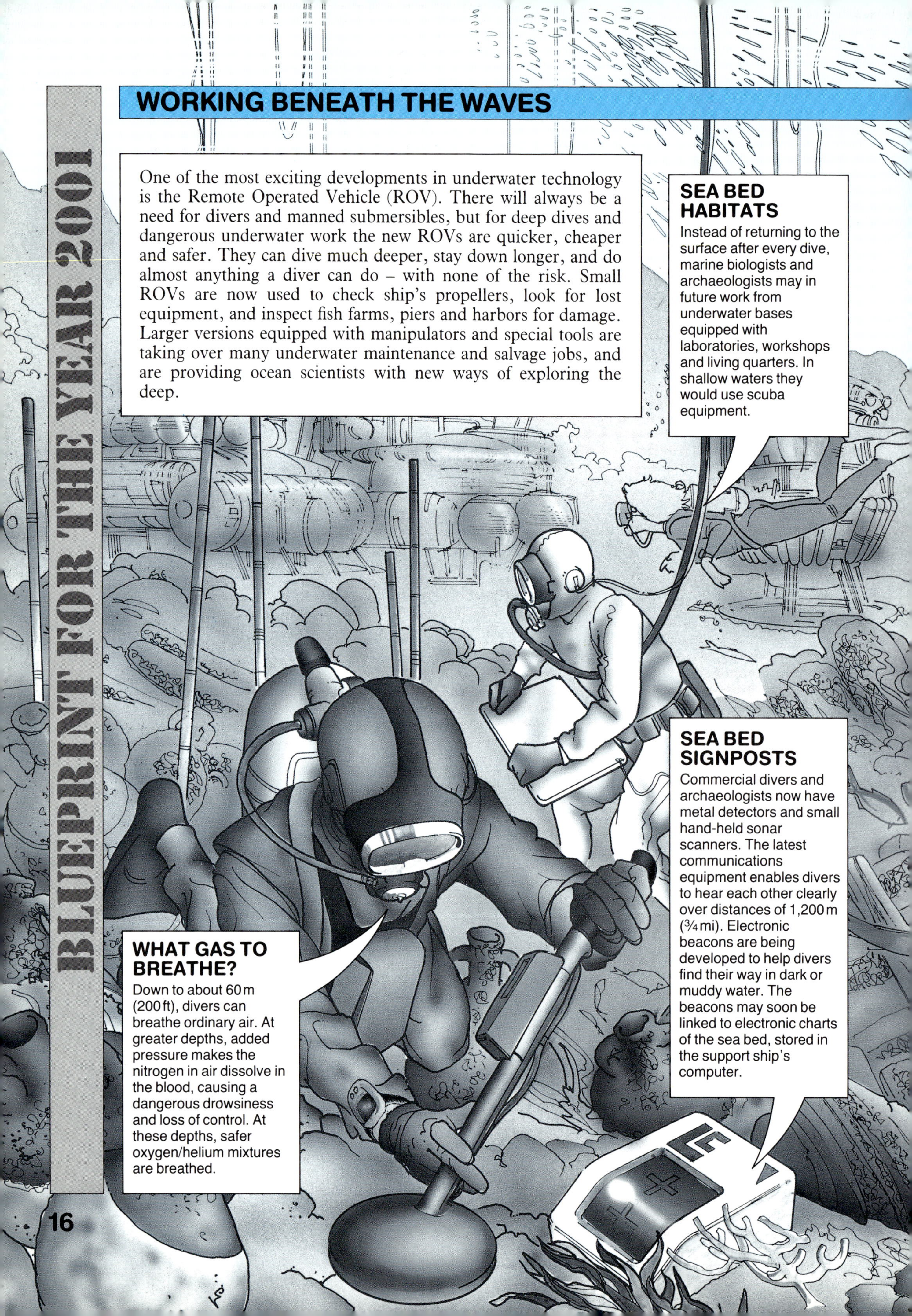

SEA BED HABITATS

Instead of returning to the surface after every dive, marine biologists and archaeologists may in future work from underwater bases equipped with laboratories, workshops and living quarters. In shallow waters they would use scuba equipment.

SEA BED SIGNPOSTS

Commercial divers and archaeologists now have metal detectors and small hand-held sonar scanners. The latest communications equipment enables divers to hear each other clearly over distances of 1,200 m (¾ mi). Electronic beacons are being developed to help divers find their way in dark or muddy water. The beacons may soon be linked to electronic charts of the sea bed, stored in the support ship's computer.

WHAT GAS TO BREATHE?

Down to about 60 m (200 ft), divers can breathe ordinary air. At greater depths, added pressure makes the nitrogen in air dissolve in the blood, causing a dangerous drowsiness and loss of control. At these depths, safer oxygen/helium mixtures are breathed.

ONE-PERSON SUBS

One-person submarines and armored diving suits are ideal for jobs requiring short dives to depths of 250–400 m (820–1300 ft). The diver remains at normal pressure throughout the dive, and this avoid the lengthy decompression procedures that make other kinds of diving slow and expensive.

WORKING SUBS

Working submersibles now range in size from mini-ROVs about the size of a portable TV set to monster machines that scoop out trenches on the sea bed and lay pipelines in them. In between are vehicles for inspection, repair and maintenance work, with attachments for cleaning, welding, cutting and even for underwater painting!

DEEP ROVER

The *Deep Rover* submersible is the work of British designer Graham Hawkes, working for Deep Ocean Engineering in California. The vehicle's chassis is aluminum, and the crew compartment is an acrylic sphere 1.7 m (5½ ft) across and 16 cm (6 in) thick.

Deep Rover's manipulator hand can be positioned to an accuracy of a fraction of an inch. Its touch-sensors produce a sound signal that tells the operator he or she is handling wood, steel, or some other material.

HUNTERS AND CATCHERS

Fishing is the only large-scale hunting for food that is carried out today. Altogether we take more than 72 million tons of fish and shell-fish from the world's oceans every year.

Most of the fishing in developing countries is carried out in shallow coastal waters, using very simple equipment. In some parts of the world – especially some of the poorest areas – fish is the only animal protein available. However, the largest share of the world catch still goes to the industrialized nations. Only they have the ships and the technology for catching fish in all depths of water and bringing them back to port in perfect condition, even after months at sea.

Commercial fishing relies on a curious mixture of ancient and modern skills. To be successful, fishing boat captains still need to have a "feeling" for where the fish are to be found. This has been so throughout the world and through the ages. But today's captains know where the fish feed, and where they breed. They have satellite navigation, radio, echo-sounders, and nets that can be aimed. Most important of all, they have automated factory ships for processing any seafood.

▽ *Wave Crest* uses a net called a purse seine. It is set in a circle around a fish school, then hauled in so that the fish are trapped in a net bag. Once alongside, the fish are pumped into the ship's hold.

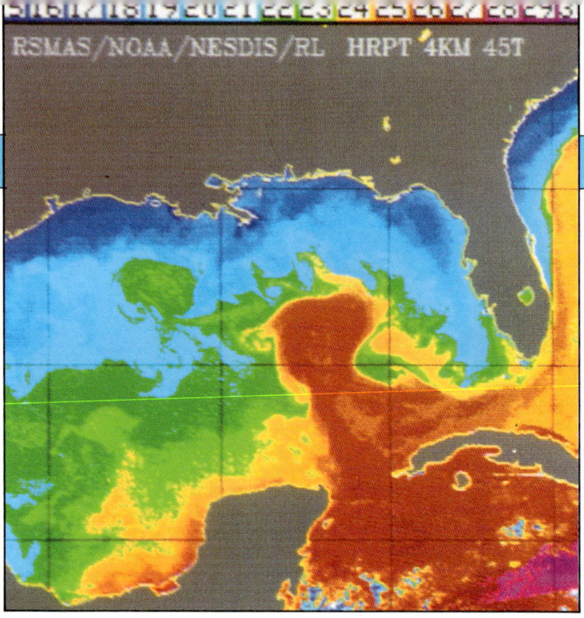

△ Computer-generated colors show temperature zones in the waters of the Gulf of Mexico.

▲ HUNTING BY SATELLITE

In 1979, the National Oceanographic and Atmospheric Administration (NOAA) launched its Coastal Zone Color Scanner (CZCS). This is a satellite designed to measure the amounts of blue and green light being reflected from the surface of the sea. The tiny green plants called phytoplankton that float in the sea contain chlorophyll, just as land plants do. This chemical absorbs blue and green light, so the more plankton there is, the less blue and green light will be reflected. The satellite can therefore detect where the water is richest in food. This is where fish are likely to gather. Other satellites measure water temperatures to reveal the cold currents and upwellings that also provide rich feeding grounds for shoals of fish.

▼ THE AIMED TRAWL

Electronics and sonar equipment have taken the guesswork out of catching fish with a trawl net. Once a shoal has been located by the forward-scanning sonar on the ship, its exact depth is measured by a downward-pointing sonar. With the position of the shoal clearly displayed on a TV screen on the bridge, the captain can steer the net straight at the fish. Echo-sounders on the net itself indicate how far above the sea bed the net is moving. By remote control, the captain can engulf the entire school in the net.

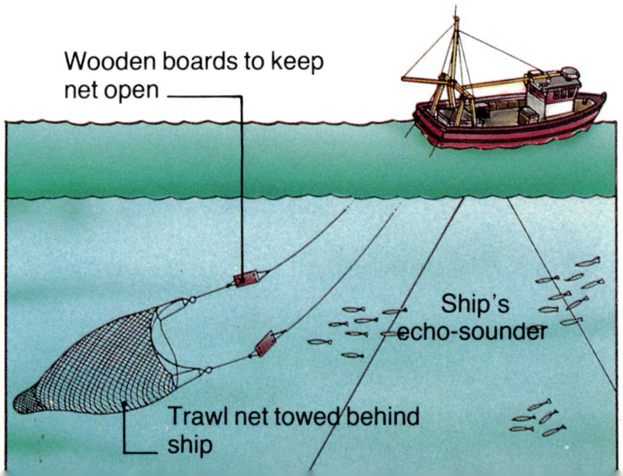

▷ The 152 crew and 368 factory workers on the *Sodruzhestvo* work two 12–hour shifts to keep the production lines running day and night.

In the spring, the ship operates in the Arctic Sea, processing the high quality crabmeat harvest. The rest of the year is spent processing fish in the north Pacific.

▲ FLOATING SUPER-FACTORY

In September 1987, the *Sodruzhestvo* was launched. This is the world's newest and most advanced factory ship. It is 180m (590ft) long, weighs 3,200 tons, and was built by the Rauma-Repola shipyard in Finland for the USSR Ministry of Fisheries.

Sodruzhestvo is the most highly automated and versatile factory ship ever designed. The engine rooms, fuel and ballast systems and entire factory operation are all under computer control.

At full capacity, the factory can process 300 tons of fish or 120 tons of crabs in each 24-hour period. The fish-canning production line is capable of turning out five cans every second. Freshly caught fish, delivered by the fleet of small catcher boats, is processed immediately into canned or frozen fish products and fish oil, while skin, bone and scraps are made into fish meal.

▼ JIGGING FOR SQUID

Squid jiggers are among the strangest-looking of all fishing boats. They are based on a fishing method that has been used for centuries in Japan, the Mediterranean and parts of Africa. But now they have been brought up to date and mechanized. Instead of hand-held lines, the modern boats have up to 24 automated line-jiggers. The jiggero pay out multi-hooked lines with plastic lures to depths of 30 to 140m (98–460ft) then wind them in again. Rectangular or egg-shaped reels make the hooks rise through the water with jerky movements that hook the squid as they try to take the bait.

▽ A modern Japanese squid jigger, like this one operating in the South Atlantic, may generate 200 kilowatts of light power to lure the squid close to the ship. The boats operate at night.

SAFEGUARDING THE STOCKS

The seas and oceans are so enormous that it is tempting to think of them as a never-ending source of food. But this is not the case. Good fishing grounds can easily be ruined by over-fishing, that is, by taking out too many fish too quickly. It leaves the fish populations with no time to recover.

Rich fishing grounds can also be ruined by the use of certain types of fishing gear. Weighted trawl nets dragged over the sea bed may catch bottom-dwelling flatfish very efficiently, but if used too often they can do immense damage to the seabed habitat.

There is no reason why fisheries should not provide a rich and constant harvest. However, this can come only with good management. The fish are "renewable" living resources. With good management what we take out can be replaced. That is why it is so important to have international agreements.

△ The water treadmill built at the Scripps Institution can be used in the laboratory or on board a research ship for studies at sea.

▼ CONTROLLING THE CATCH

If fish stocks are to remain stable and provide a continuous harvest, it is essential that the amount of fish caught be carefully controlled. This is done by setting limits on the total tonage that can be landed at fishing ports.

Another important way of avoiding over-fishing is to control the size of the fish that are taken. It is best to catch only the larger, older fish because they will already have spawned and produced young that can replace them. The younger and smaller fish must be allowed to escape so that they, too, can grow to be adults and spawn. This is done by regulating the size of mesh used in nets, so that the smaller fish can slip through.

▲ FISHERIES RESEARCH

The more scientists can find out about the way fish live and the way they behave, the more we can do to use the ocean food resource efficiently and wisely. Effective conservation programs depend first of all on basic scientific research.

Dr. Jeffrey Graham, a marine researcher at the Scripps Institution of Oceanography in La Jolla, California, has recently developed equipment that allows biologists to investigate the swimming behavior of large fish. The device is like a water treadmill. The fish is placed in a tank, swimming against a current of water pumped through the tank, while the scientists study its muscle movements, heart beat and breathing.

▷ A Canadian fishery patrol aboard a foreign vessel. Most of the world's fishing nations now have strict regulations to control fishing in their national waters. Fishery patrols can board any boat fishing in their area and see that it is using the approved size of mesh. Fishing boat captains who break the rules may be arrested and have all their gear taken away.

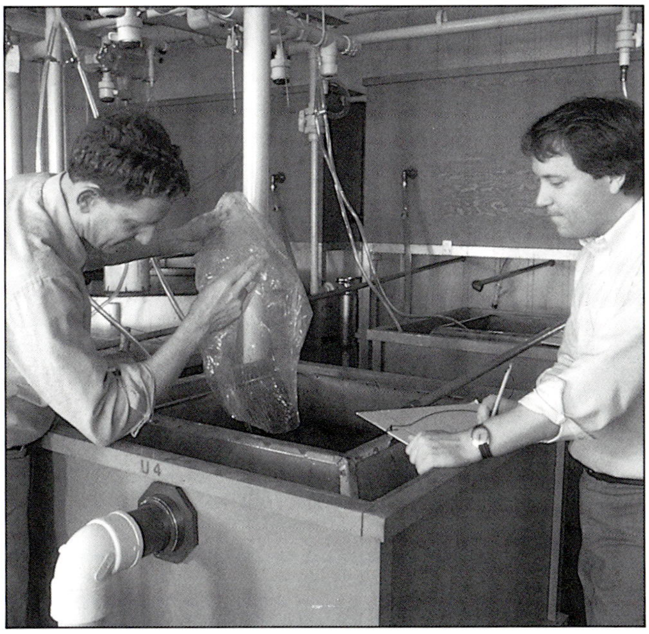

◉ FISH FARMING

People have dreamed of maritime fish farming for hundreds of years. Now the idea is becoming a reality. Fish can be produced in large numbers in artificial scientifically controlled conditions, and they can be used in two ways. They can be marketed directly for human consumption, or they can be released into rivers or coastal waters to boost natural populations that have become smaller.

In Maryland, the Baltimore Gas and Electric Company has come up with a novel idea. The company has sponsored a fish farming system that uses waste heat from one of its coal-fired power stations, in the form of hot water, to raise striped bass. These are later released into Chesapeake Bay where striped bass are becoming scarce.

◁ ▽ 200,000 newly hatched striped bass are placed in each warm water tank (left). The diagram (below) shows the layout of the striped bass "farm" described above.

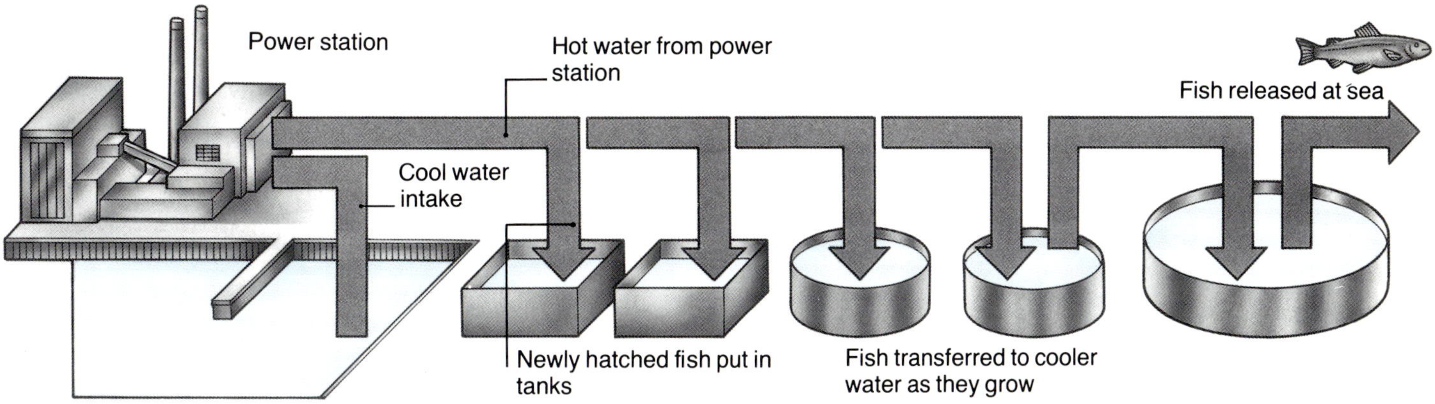

◉ PONDS AND LANTERN NETS

Many different fish farming methods are used in Japan. Fresh water species such as trout, carp, eel and a new species called the *ayu* are raised in tanks and in artificial ponds. Marine fish such as yellow-tail jack and red sea bream are raised in large cages floating in sheltered bays around the coast. Since oxygen-rich water passes through the net cages all the time, huge numbers of fish can be kept in each cage.

Japan, however, is best-known as a leader in the farming of shellfish, especially oysters and scallops. The scallop larvae, or spat, are collected at sea, usually by hanging net collecting-bags full of cedar twigs in the water. The spat are grown in seawater tanks and then transferred to "lantern nets" that hang in long lines from floating buoys along the coast.

◁ Scientists at St.Andrew's biological station in Canada are investigating methods of growing scallops in lantern nets as shown here.

SEA FOODS OF THE FUTURE

Although it is unlikely that any major new fishing grounds will now be discovered, there are several ways in which we can increase the amount and variety of food we get from the sea.

According to the United Nations Food and Agriculture Organization (FAO) almost 80 per-cent of the world fish catch comes from the northwest and southeast Pacific and the northern and eastern Atlantic. Yet while these areas are being fished to the limit, other areas are under-used. Future fisheries will make better use of these areas.

Another problem is fashion in food buying. Many perfectly good food-fish are currently caught but not used because they are ugly or bony, and people will not buy them. New ways of processing the flesh of such fish into more acceptable foods will avoid this wastefulness.

△ Japan has a long history of seaweed cultivation. The red seaweed *Nori*, and the brown seaweeds *Wakame* and *Konbu* are high value food crops, while kelp (above) is farmed for its chemicals.

▼ THE FOOD-AND-FUEL PLANT

Seaweed is one of the growth industries of the marine world today, and attention has now focused on the giant Californian kelp, *Macrocystis*, which can grow 60cm (24in) a day and reach 60m (200ft) in length.

On an experimental test farm 8km (5mi) off the coast of California, the Gas Research Institute of America is growing kelp on a vast spider web of steel cable surrounding a central processing plant. The kelp is fed by nutrient-rich water pumped from far below the surface, and is harvested by special ships equipped with cutters that are like giant hair clippers. The weed is processed into methane gas for fuel, animal feed, and a variety of chemicals. Of these, the most important are called alginates. They are used in products as sausages and paper. In a second set of experiments, kelp plants are being grown in floating plastic "salad bowls."

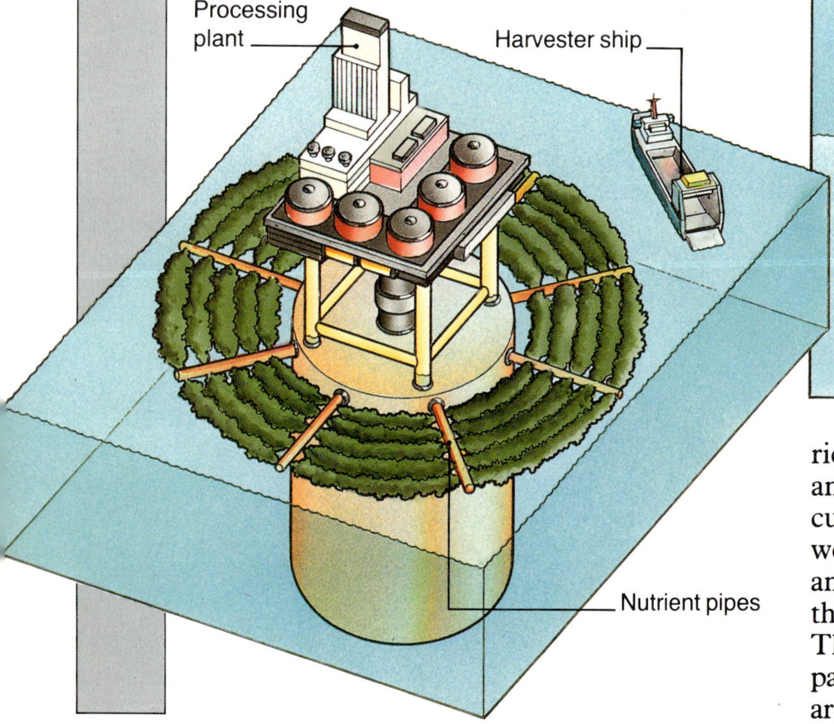

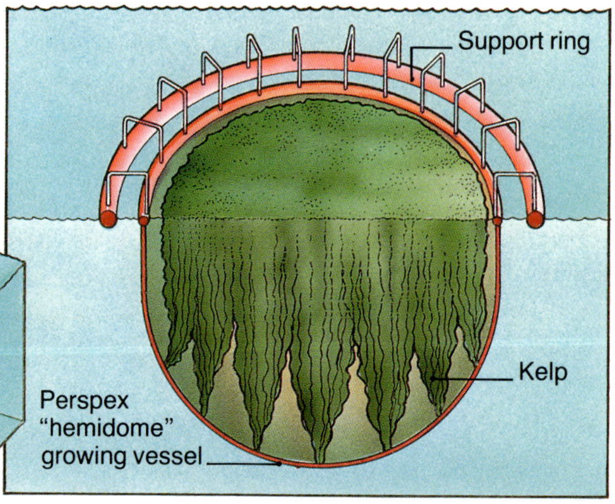

THE SOUTHERN HARVEST

The cold waters of the southern oceans hold an enormous stock of seafood that was completely overlooked until the mid-1970s. This resource consists of the Antarctic krill, the minute shrimp-like animal that is the main food of baleen whales.

Krill can be fished using existing fishing boats. The tiny animals form enormous schools hundreds of feet across, so that a large fine-mesh trawl net towed through the swarm can fill with up to 60 tons of seafood in a single haul. Already, the USSR, Japan and Poland have full scale krill fishing operations in the southern oceans, taking several hundred thousand tons of krill each year.

Krill is easy to catch but less easy to handle and store. It is usually processed into a protein paste and then used as a food additive or molded into flavored fish sticks. Fresh krill tails are sold as a delicacy in Japan. The meat is very rich in protein and contains most vitamins. However, as demand for krill rises, our use of the resource must then be carefully studied and controlled, because seals, whales, birds and fish all depend on this source of food.

△ Just how much krill there is in the cold southern seas is very difficult to assess: current estimates vary from 50 million to 2,000 million tons! Whatever the true figure, it is an enormously rich food resource.

NEW FISH – NEW PRODUCTS

Many perfectly good food-fish are thrown back into the sea, or dried and turned into fish meal for use in animal feed or fertilizer, simply because people find them unattractive or bony and refuse to buy them.

One solution to this problem may lie in current research into fish processing methods and machines. For example, scientists at the Torry Research Station are working on filleting machines that will handle tough-scaled or bony fish such as scad and blue whiting. The Torry laboratory was also one of the leaders in Europe in the development of fish paste and surimi. Surimi is a colorless protein gel made by repeatedly washing pressing boned fish meat. It can be made from any of the unwanted fish and molded into almost any seafood.

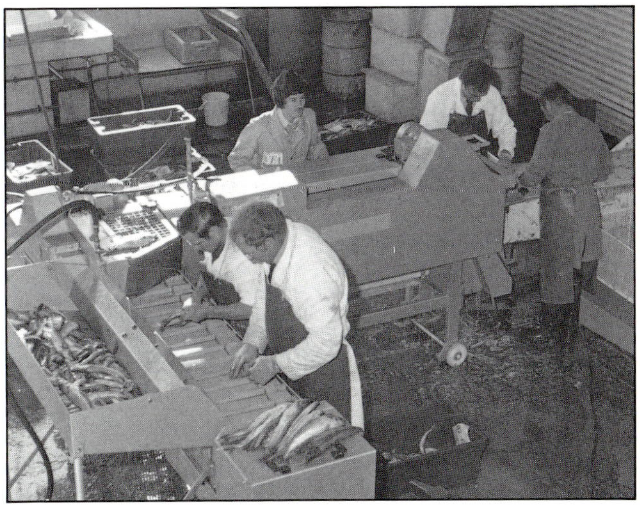

△ Fish-processing machinery at the Torry laboratory.

◁ Some of the less popular fish that can be eaten.

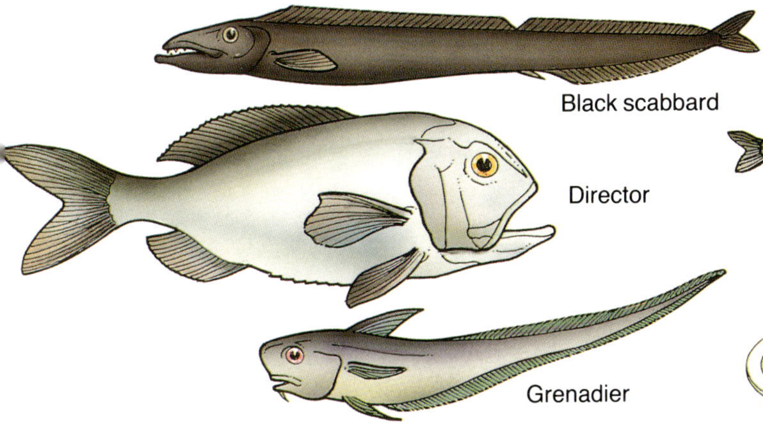

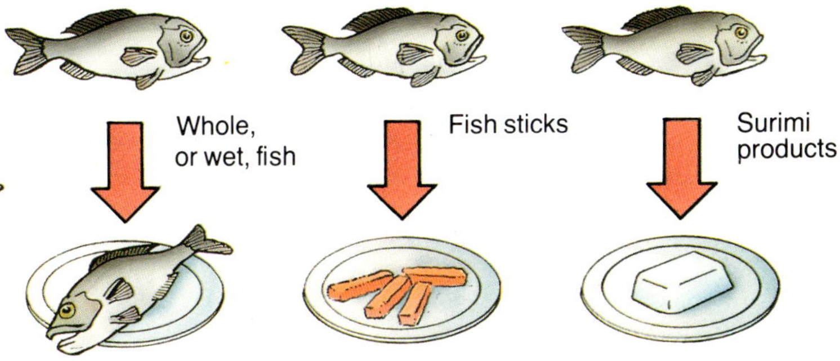

THE OCEAN HARVEST

Great changes are taking place in *where* we fish, *how* we fish and what we *do* with the fish we catch. In the next 10 years, better use will be made of currently under-used parts of the oceans. Advances in satellite tracking, sonar equipment and net design will improve the efficiency of the catcher boats. New technologies will enable us to make use of the krill harvest and of many fish species that have been wasted in the past. Altogether these changes could increase the total world catch of fish and seafood from just over 70 million tons to 120 million or more. Ships will be unloaded by computer-controlled pumps and conveyors, and the catch will be sorted and processed in automated factories.

SEAWEED POWER

Seaweed harvested from natural kelp beds or from offshore seaweed farms may be processed to produce synthetic natural gas. (SNG). This may then be burned to generate electricity. Scientists estimate that a kelp farm covering 2.5 sq km (1 sq mi) could provide all the energy needed by up to 2,000 people.

AUTOMATION

Wharves of the future will be almost entirely automated. Fish, kelp and other sea produce will be transferred from ship to shore by pumps, conveyors and robots, all under the control of highly trained food-processing engineers.

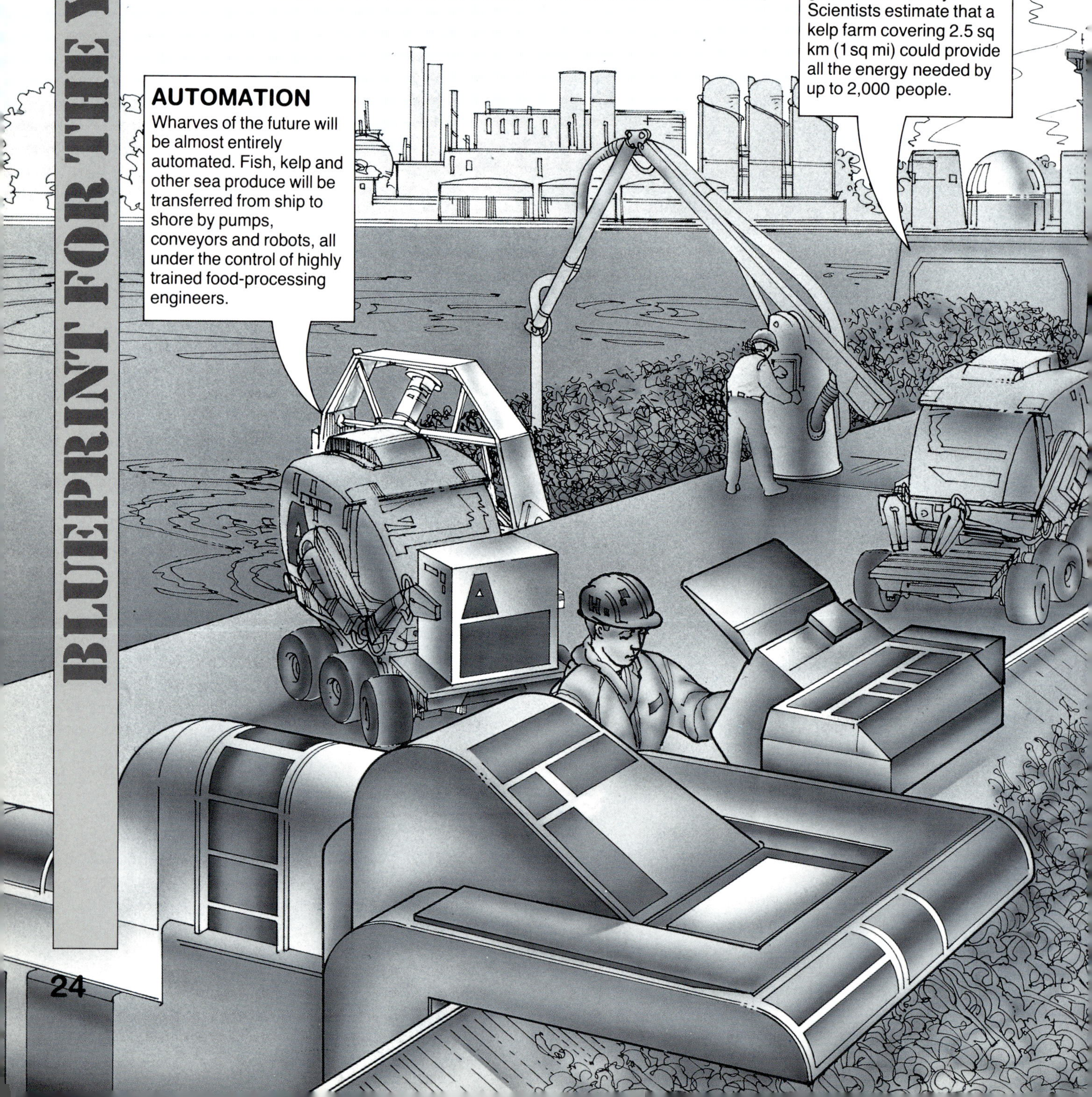

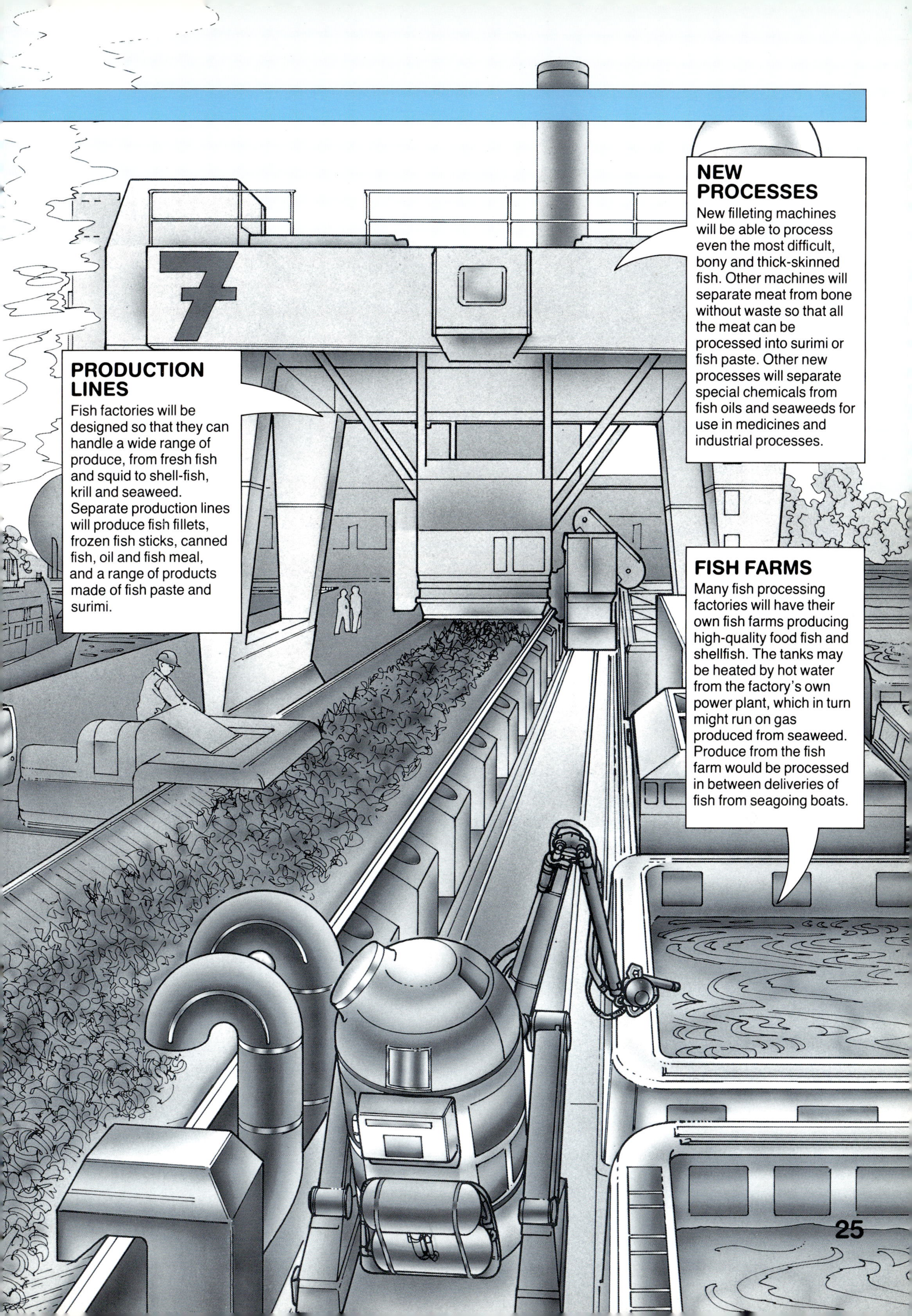

PRODUCTION LINES

Fish factories will be designed so that they can handle a wide range of produce, from fresh fish and squid to shell-fish, krill and seaweed. Separate production lines will produce fish fillets, frozen fish sticks, canned fish, oil and fish meal, and a range of products made of fish paste and surimi.

NEW PROCESSES

New filleting machines will be able to process even the most difficult, bony and thick-skinned fish. Other machines will separate meat from bone without waste so that all the meat can be processed into surimi or fish paste. Other new processes will separate special chemicals from fish oils and seaweeds for use in medicines and industrial processes.

FISH FARMS

Many fish processing factories will have their own fish farms producing high-quality food fish and shellfish. The tanks may be heated by hot water from the factory's own power plant, which in turn might run on gas produced from seaweed. Produce from the fish farm would be processed in between deliveries of fish from seagoing boats.

ENERGY FROM THE SEA

The first great rush of oil from the world's shallow-water fields is now almost over, and the energy companies are supporting two principal lines of research for the future. The first is to squeeze as much oil as possible from existing fields. The second is to build seabed production systems that will eventually replace the huge platforms used at present. These giant structures of steel and concrete are very expensive to build and run; a medium-sized platform in the North Sea, for example, towers 225 m (750 ft) from the sea bed to the top of the drilling rig, supports more than 22,000 tons of equipment, and accommodates 200 people. The new seabed manifolds will be remote controlled and serviced by robots. They will be cheaper, safer, and will pave the way for the move into deeper waters.

At the same time, research continues into ways of harnessing the energy resources of waves and tides, and the heat energy stored in the oceans.

▶ OFFSHORE OIL AND GAS

One way of extending the life of an oilfield is to pump water or waste gas down some of the bore-holes. This increases the pressure in the deep rock layers and forces out more oil. It works well, but produces oil mixed with water. An idea being explored by French and Norwegian engineers is to pass the mixture of oil, water and gas through a machine like a huge food processor on the seabed. This would whip the mixture into a creamy foam which could then be pumped ashore and easily separated.

Smaller oilfields do not have to have their own platforms. They can be linked by seabed pipelines to a central platform. Another solution is the storage system installed by Shell in 120 m (400 ft) of water off the coast of Spain. Oil rises from the sea bed through a swivel-jointed pipe to an oil tanker permanently moored over the well. The tanker is equipped to process the crude oil and then off-load to smaller tankers.

Most subsea oil wells of the future will be controlled by Underwater Manifold Centers (UMCs) instead of production platforms. (The manifolds are bore-hole covers with pipes bearing valves.) They are built on a standardized part system so that valves, control units and production units can be replaced by a robot.

▷ At first, UMCs will be used alongside existing production platforms, but over the next 10 years the manned platforms will probably be phased out.

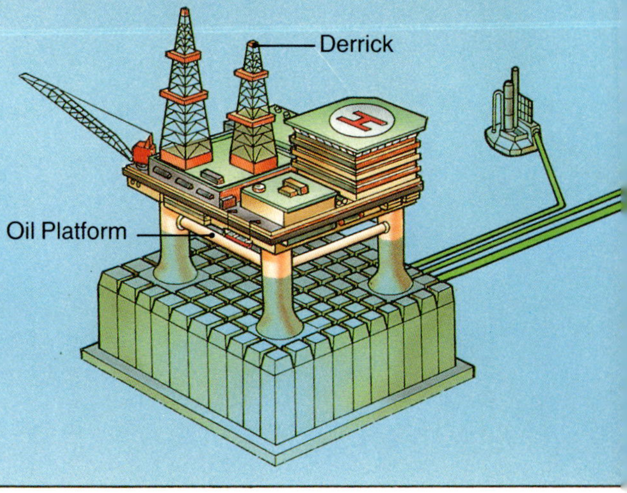

▷ To develop new oil and gas fields in the North Sea will mean working in up to 350 m (1,000 ft) of water. Experimental designs for deep-water platforms include a steel tower on a huge swivel joint, and a slim concrete tower held down by steel piles.

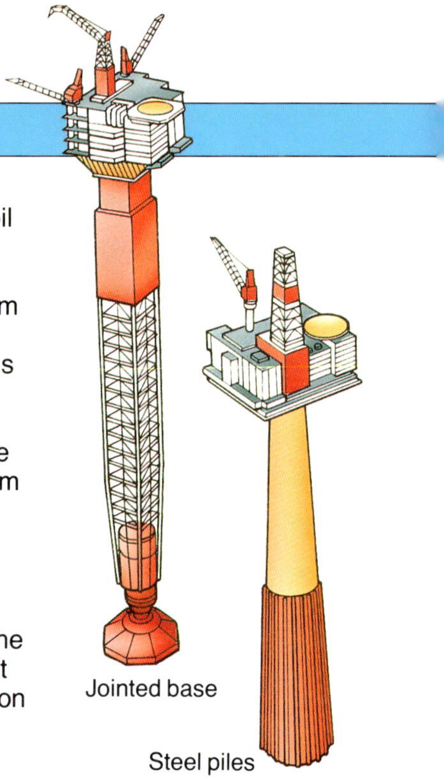

▽ The Underwater Manifold Center for the North Sea Cormorant Field under construction at Schiedam in the Netherlands.

▶ OTEC

The simple fact that tropical seas are much warmer at the surface than they are deep down offers engineers the chance for unlimited free energy. The idea of OTEC (Ocean Thermal Energy Conversion) has been around for several years, but in 1979 the first trial plant was run successfully near Hawaii. It was a small plant, generating just 50kW of electricity, but it proved the theory worked. Now there are plans for OTEC plants 8,000 times the size of the test plant.

How does it work? Warm water enters through grilles in the top and its warmth is used to evaporate liquid ammonia, that is, convert it from liquid to gas. As it changes, the ammonia expands violently and this force is used to drive the generator turbines. Once its job is done, the gas flows into tanks cooled by much colder water pumped up from the deep. It immediately turns back to liquid, and can be used again. The ammonia goes round and round. All that is needed to drive the plant is warm and cool water with a temperature difference of at least 20°C (68°F).

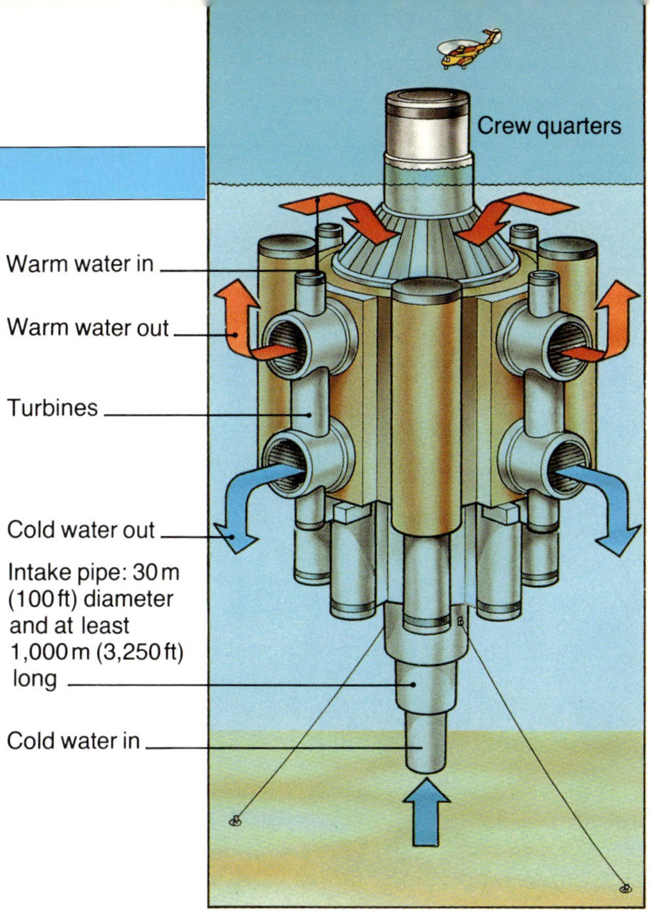

▶ WAVE POWER IN NORWAY

Like many brilliant ideas, the Oscillating Water Column is basically simple. (To oscillate means to move to and fro.) Waves rushing up to the foot of a cliff are channeled into the base of a concrete tower. As the water surges up inside the tower, it acts like a powerful ram or piston. The air in the tower is forced upward and out through the blades of a turbine at the top. As the wave sweeps out again, air is sucked back into the tower, spinning the turbine blades again. The wave piston idea was developed by the Kvaerner Company in Oslo, the research was supported by the Norwegian Government, and the special turbine, which spins the same way no matter which way the air flows, was designed by Professor Wells at Queen's University in Northern Ireland.

▽ Kvaerner's prototype wave power generator was installed in 1985. More are being built. They could be especially useful in remote areas.

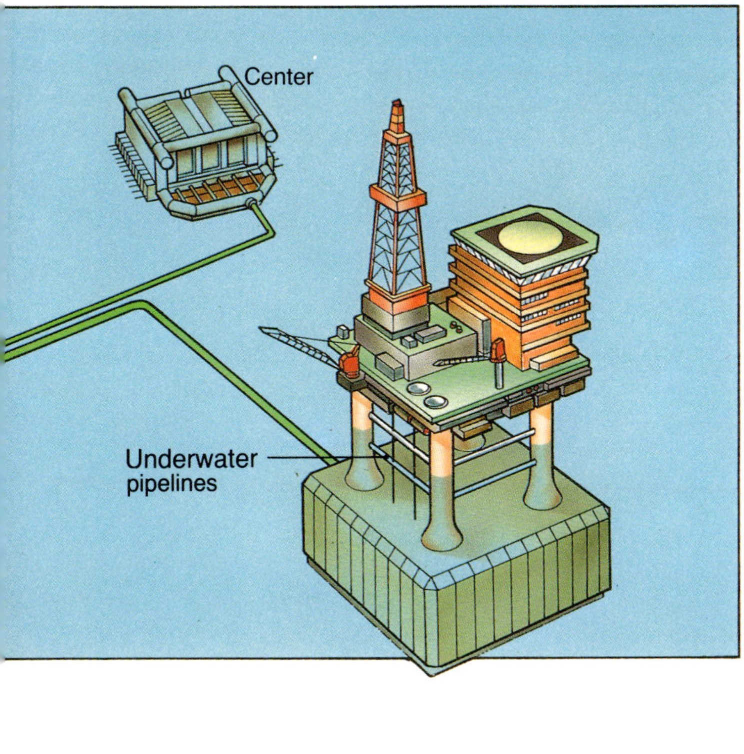

WATER AND MINERALS

One of the most important properties of water is that it will dissolve more chemical substances than any other liquid. As a result, the oceans are a vast storehouse of minerals.

Seawater is 96.5 percent water and 3.5 percent dissolved chemicals. The two most common are sodium (1.0 percent) and chloride (1.9), which combine to form salt. The percentages may seem small, but there is enough salt in the oceans to cover the land with a layer 150 m (500 ft) thick. Magnesium, calcium, sulphur and potassium are the next most common elements. About 80 others are also present, but in minute quantities. Metals such as iron, manganese, zinc and copper settle out of the water as tiny solid particles.

DUAL PURPOSE ICEBERGS

Water itself is one of the world's most precious resources, and our freshwater supply is replenished constantly by water evaporated from the sea.

Some parts of the world are desperately short of freshwater. The Arab countries, western Australia and the dry Southwest of the United States are rapidly using up what underground supplies they have. The idea of towing icebergs from Antarctica may soon become a reality. Towing a berg is not a problem. Already, Arctic bergs are regularly towed clear of the oil rigs and shipping lanes off the Labrador coast. The new idea is to use the berg as a power source as well as a source of freshwater.

DESALINATION

One of the most difficult things to separate from seawater is salt. A major problem is that the salts in seawater are corrosive. Desalination processes—the removal of these salts – involve costly metals. A widely used method is called flash distillation. The seawater is heated, then passed into a low-pressure chamber. The sudden drop in pressure makes some of the water "flash" into steam. This is then condensed back into liquid by passing it over pipes full of cold water.

One of the world's biggest desalination programs is in Saudi Arabia, where plants along the Red Sea and Gulf coasts produce millions of gallons of freshwater a day.

△ A desalination plant producing drinking water for the Caribbean island of Bonaire.

The ice-power idea uses the same scientific principle as OTEC (page 27). A captured iceberg is broken up and the pieces are towed into a freshwater lagoon. As they melt they provide freshwater, but at the same time the very cold water of the lagoon is used as the cooling liquid in a generator based on evaporating and condensing ammonia.

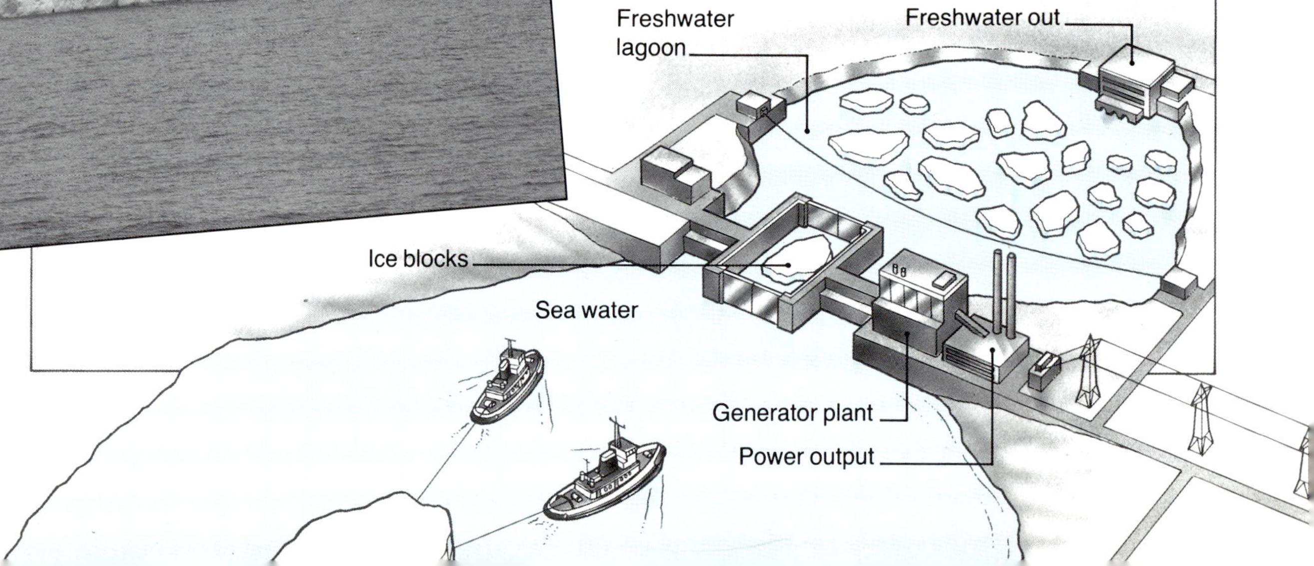

▶ MINERAL NODULES AND CRUSTS

Potato-shaped lumps, or nodules, rich in minerals were discovered on the deep seabed by the Scripps Institution of Oceanography in the 1950s. Since then they have been studied and mapped carefully. In cross-section, the nodules are made up of layers, like an onion. Dark, iron-rich layers alternate with lighter layers rich in manganese. Even more important, in some areas the nodules contain useful amounts of very valuable minerals such as cobalt, nickel and copper. Soon, it will be economical to mine then.

One clue to how the nodules form is that they seem most common in areas where there is a lot of microscopic animal life in the water. Many sea creatures collect and store minerals in their bodies. When they die and sink to the sea bed, they enrich the mud at the bottom with the minerals and thus help create the nodules. Very recently, mineral-rich crusts over an inch thick have been found on some seabed rocks. The crusts are similar to nodules in composition but are even richer in cobalt.

▽ Hot black liquid gushes from a "smoker" on the bed of the East Pacific Rise. Smokers were first studied using the manned submersibles *Alvin* (U.S.) and *Cyana* (France).

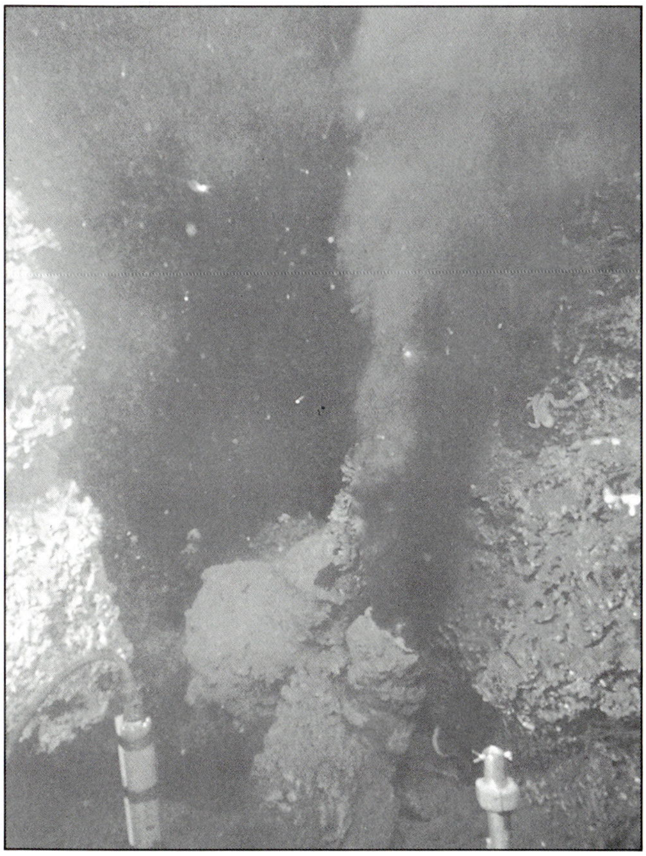

▷ Manganese nodules lying on the sea bed.

▽ Seabed mining machines include a suction dredge like a deep-sea vacuum cleaner tested in the United States, and the tracked miner, here, designed by West German companies for mining nodules in the Red Sea.

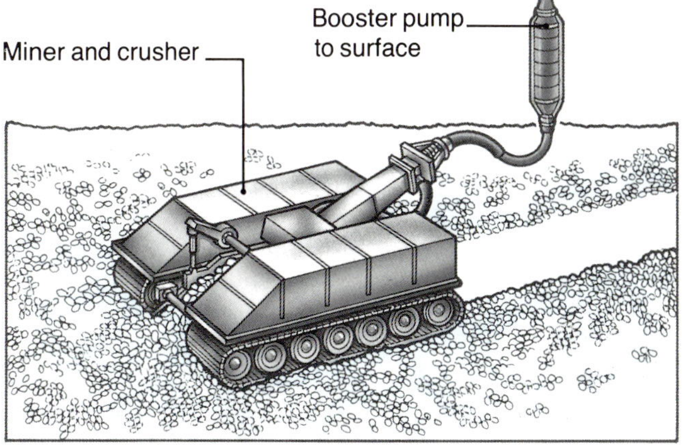

▼ SEABED "SMOKERS"

Between 1974 and 1984, French and American scientists working in the Pacific made a series of exciting discoveries. They were studying the mid-ocean ridges of the eastern Pacific, where new ocean crust is being formed by molten rock rising from below. In parts of the ridge system, temperatures one or two miles beneath the sea bed may be as high as 1,400°C (2,500°F).

The scientists were investigating these geological processes in action using submersibles and camera sledges. What they also discovered was a new kind of sea floor biology (see page 31), and chimney-like vents in the sea bed that were pouring out clouds of dense black liquid at 350°C (650°F).

▷ As the hot fluid pours out, iron sulphide forms a thick black cloud. The chimney is made of a mixture of sulphates and sulphides of iron, zinc, copper and silver. All are commerically valuable.

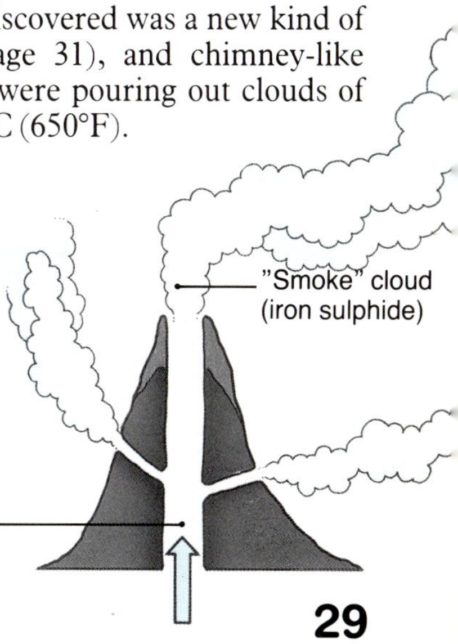

MARINE BIOTECHNOLOGY

Marine biology today is poised on the brink of a revolution. Over the next 10 or 20 years, this branch of science could completely change the lives of millions of people throughout the world. The reason for all the excitement is that the new sciences of biotechnology and genetic engineering are now turning their attention to the oceans—with dramatic results.

New chemical compounds are being discovered almost every week, in seaweeds, fish oils, and such unlikely animals as sponges, sea squirts, corals and worms. Many of these chemicals show great promise as sources of new drugs for use against cancers, leukemia, heart disease and other disorders. Other compounds will almost certainly provide new pesticides.

Genetic engineering may, in the future, provide new ways of greatly increasing the amount of food that can be produced by fishfarming methods. For example, it may improve the size or breeding of selected food species.

▼ PREHISTORIC LIFE SAVER

The horseshoe crab has remained unchanged for the last 140 million years, yet this living fossil has just given medical science a new weapon.

Biologists at the Marine Biological Laboratory in Woods Hole have found that a compound called LAL, made from the crab's blood, can detect minute amounts of certain poisons produced by bacteria. The poisons, called pyrogens, can cause fever, shock, and even death in humans, yet they were getting into medical drugs and other fluids meant for injection. The problem was that pyrogens can remain in water after all bacteria have been killed, even in the carefully purified water used in drug manufacturing.

△ The clam worm or pile worm of north California is one of the large *Nereis* worm family. They are large, free-swimming hunters.

▲ CHEMICAL WEAPONS

Fishermen have know for years that flies die if they land on the dead remains of certain marine worms. The reason for this became clear when it was discovered that one particular kind of worm contained a powerful poison. Once this poison had been analyzed, it was used as the base for an effective insecticide. The great advantage of the *Nereis* poison is that it kills rice stem borers and other insect pests but does not harm warm-blooded animals.

Many marine animals contain poisonous substances in their bodies. Several of the poisons are unlike any chemicals found on land, so they are of interest to biochemists. Some may have medical uses. Others may provide a basis for new weapons against crop-destroying pests. Some may be used just as they are, while others could be used for new synthetic chemicals.

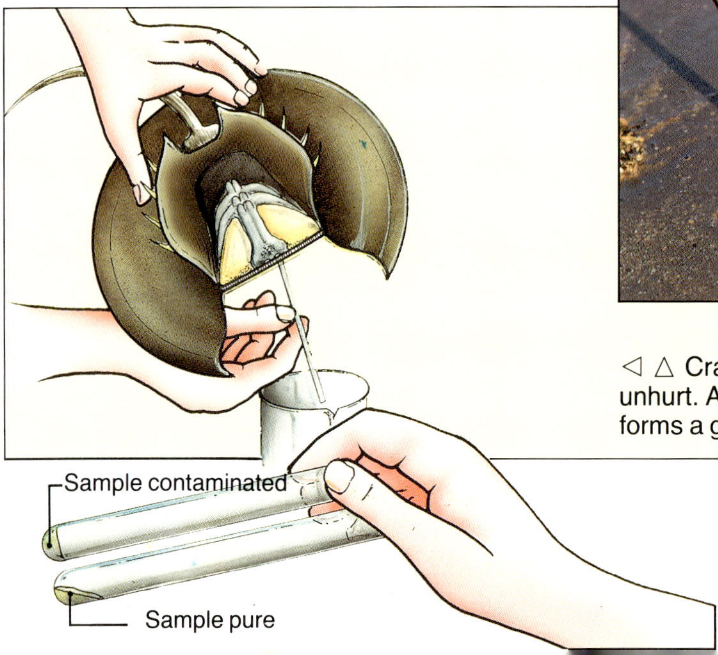

◁ △ Crabs are bled, and returned to the sea unhurt. A test sample is mixed with LAL. If it forms a gel, the sample contains pyrogens.

▶ GENETIC ENGINEERING

The marine plants and animals that have already been picked out as potentially useful to us include members of many different families (*right*). Possible anti-cancer chemicals have been found in Caribbean corals. A sea squirt from the same region contains chemicals that slow down the growth of certain viruses. Several sponges contain substances that may combat heart disease, and insulin for treating diabetes can be extracted from tuna fish.

Perhaps even more exciting is the new science of genetic engineering. The blueprint for any animal or plant is carried in a chemical code inside the nucleus, or control center, of every cell. Inside the nucleus are microscopic thread-like chromosomes, and on each chromosome there are many packets of information, called genes, written in the structure of a chemical called DNA. Each gene carries the code for a particular feature. In a human, one might be for height, another for color of eyes, and so on. In genetic engineering, small bits of the gene code are altered. Sections can be cut out and new bits spliced in, so that a plant or animal can be tailor-made for a particular use. The science is still new, but if future work on fish and shellfish goes as well as the earlier work on crop plants, genetic engineering could soon make a huge contribution to feeding people.

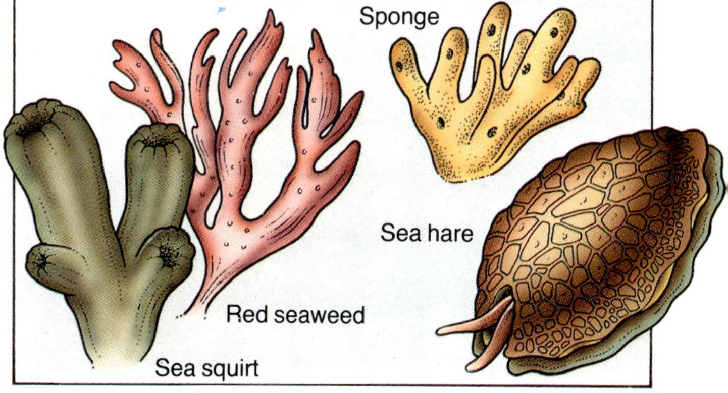

▼ CHEMOSYNTHESIS

When biologists John Corliss and John Edmond surfaced from a 2,500m (8,000ft) *Alvin* dive on the Galapagos Rift in 1977, the specimens they brought back staggered the scientific world. Around the sulfur-rich hot water springs on the seabed they had found dense clusters of animal life.

It was the worms in particular that intrigued the scientists. They were unlike any known species. They have no mouth, or gut, or anus. At the head end, a thick feathery gill plume sways in the water, enabling the worm to absorb dissolved chemicals. But it was the worms' *insides* that provided the biggest surprise of all. The body tissues are packed with specialized bacteria, and it was later found that some of these "lodgers" convert sulphide chemicals from the water into energy. They then use the energy to build carbon compounds that feed both the bacteria and the worm.

Up to this time, biologists had believed that all life depended on the energy of the Sun, converted into sugars by photosynthesis—the chemical process that takes place in green leaves. All animals either ate plants or ate other animals that ate plants. But the deep ocean communities' source of food energy was the chemical action of their microscopic partners.

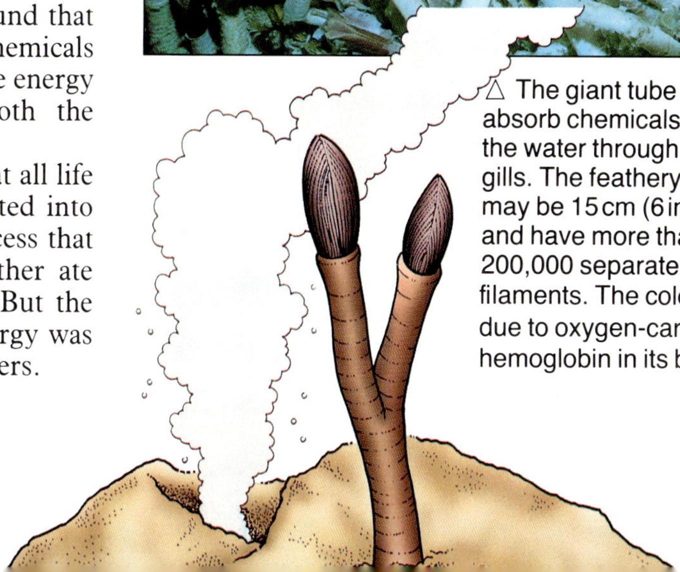

△ The giant tube worms absorb chemicals from the water through their gills. The feathery plume may be 15cm (6in) long and have more than 200,000 separate filaments. The color is due to oxygen-carrying hemoglobin in its blood.

SHARING THE OCEAN RESOURCE

The final years of the 20th century really should be a time for scientists, technologists and politicians to start planning how to make the best use of the oceans' resources. We are now in an age of computers, robotics, lasers and fiber optics, satellites and genetic engineering. The amount of information that engineers and designers now have at their fingertips should make it possible for everyone to benefit from the oceans. Wave power generators and wind-assisted cargo ships could be valuable developments in the temperate regions, while OTEC power plants, marine biochemicals and the mineral nodules that litter the sea bed could raise living standards in the tropics.

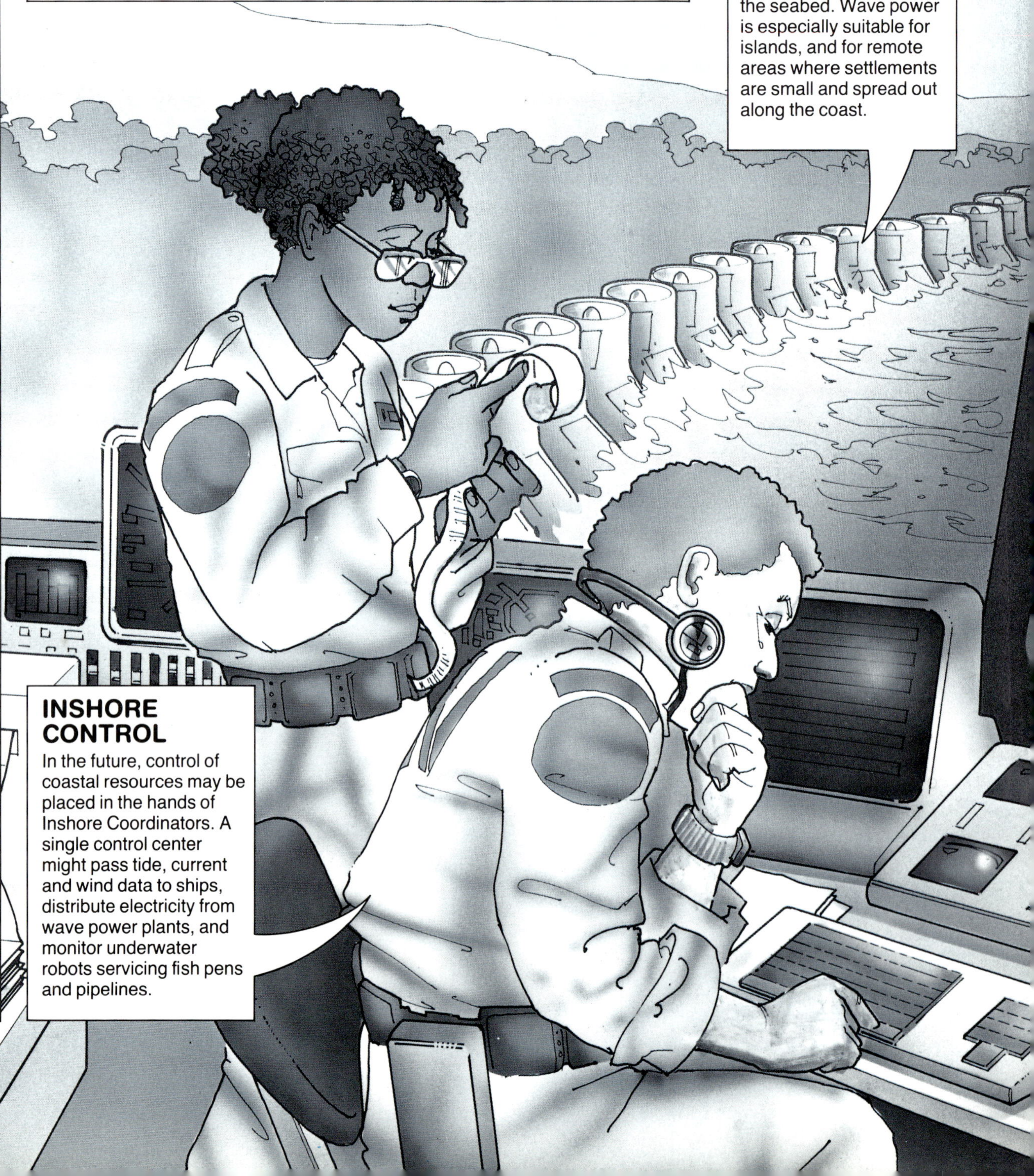

WAVE POWER 1

Wave piston generators could be built wherever there is a regular flow of incoming waves. They are easy to build on cliff sites, but could also be built in open bays, either floating or built directly on the seabed. Wave power is especially suitable for islands, and for remote areas where settlements are small and spread out along the coast.

INSHORE CONTROL

In the future, control of coastal resources may be placed in the hands of Inshore Coordinators. A single control center might pass tide, current and wind data to ships, distribute electricity from wave power plants, and monitor underwater robots servicing fish pens and pipelines.

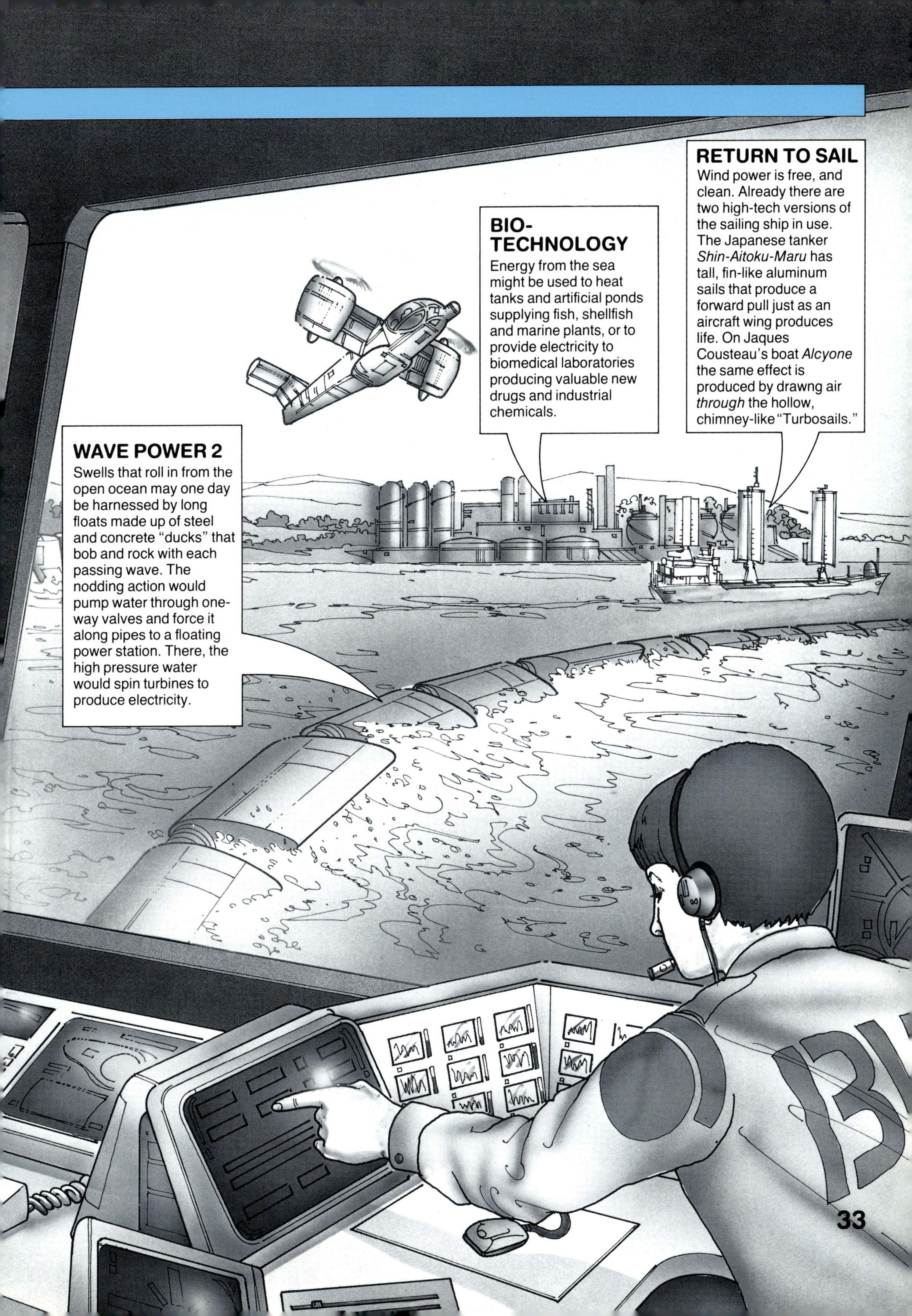

RETURN TO SAIL

Wind power is free, and clean. Already there are two high-tech versions of the sailing ship in use. The Japanese tanker *Shin-Aitoku-Maru* has tall, fin-like aluminum sails that produce a forward pull just as an aircraft wing produces life. On Jaques Cousteau's boat *Alcyone* the same effect is produced by drawng air *through* the hollow, chimney-like "Turbosails."

BIO-TECHNOLOGY

Energy from the sea might be used to heat tanks and artificial ponds supplying fish, shellfish and marine plants, or to provide electricity to biomedical laboratories producing valuable new drugs and industrial chemicals.

WAVE POWER 2

Swells that roll in from the open ocean may one day be harnessed by long floats made up of steel and concrete "ducks" that bob and rock with each passing wave. The nodding action would pump water through one-way valves and force it along pipes to a floating power station. There, the high pressure water would spin turbines to produce electricity.

CLEANING UP OUR ACT

One thing we now know for certain is that we cannot simply throw our rubbish into the sea and forget about it. Some time, somewhere, someone will end up paying the price.

It is now an offense for oil tanker operators to wash out their ships' tanks at sea. However, accidental spills still happen. Improved standards of seamanship, better navigation systems and better traffic control in crowded waterways are all helping to reduce the chances of disastrous oil (and chemical) spills. At the same time, new methods of cleaning up are being invented.

International agreements are slowly beginning to tackle the problems of waste-dumping at sea. Even our most useful farm chemicals must be treated with care. Pesticides sprayed on to the land are washed into rivers and then into the sea. Some, such as DDT, have now been banned in many countries because of their harmful long-term effects on all types of marine life.

△ Seabirds are often the first victims of oil spills. Thousands of birds died when *Amoco Cadiz* ran aground off Brittany in 1978.

COPING WITH OIL SPILLS

Oil spills cause enormous harm to coastal habitats. Animal life is killed, local fisheries are damaged, and the tourist industry may be ruined for years. Not surprisingly, oil companies and other marine specialists invest a great deal of money and time in looking for new ways of dealing with spills.

"Rigidoil" is a new system invented by British Petroleum. Two liquid chemicals are sprayed on to the floating oil, using high pressure hoses to make sure the oil and chemicals are well mixed. Within minutes, the liquid oil is changed into a clean, rubbery, solid sheet. The oil can spread no further, it can no longer harm marine life, and it can be removed easily with nets or rakes. The solidified oil can be used as land fill.

"Sea Skimmers," built by Vikoma International, can pick up floating oil at up to 100 tons an hour. The oil clings to the rotating vertical disks along the sides of the machine. It is then scraped off by plastic blades on the inside of the raft. The oil slides down into the central well, and is then pumped out to the waiting barge.

◁ Spraying on the chemicals.

▽ Picking up the solidified oil.

▽ The "Sea Skimmer" family ranges from machines that can pick up 9 tons an hour and are used to clean up harbors, to other units that can handle 100 tons an hour and can work offshore in rough weather.

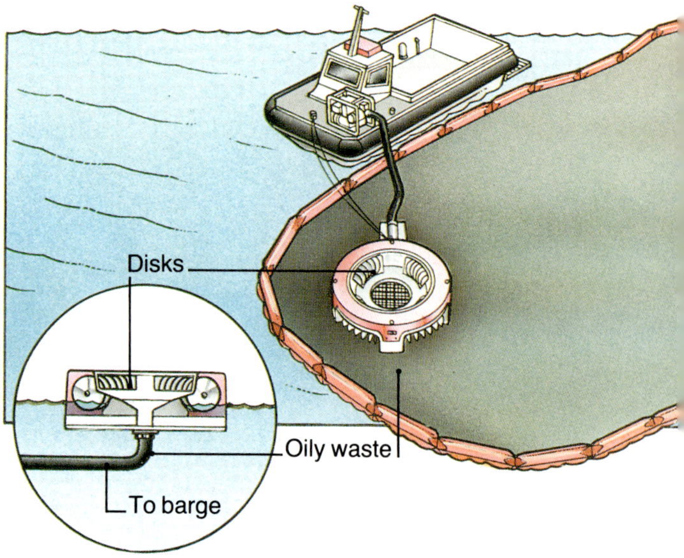

TRAFFIC CONTROL

In just two weeks, the crowded waters around northwestern Europe carry more than 25,000 shiploads. Most of the tankers and cargo vessels travel along the Channel and up the North Sea, but there are also more than 300 ship, hydrofoil and hovercraft ferry routes.

In situations like this, good traffic control is vital. Up to 500 ships a day pass through the narrow Straits of Dover, so for safety there is a one-way system. Incoming vessels take the French side; outward-bound vessels take the English side. Unfortunately not everyone follows the rules, so French and British coast guards must maintain a 24-hour radar watch to guard against collisions.

Computer graphics and high-speed handling of data are making busy waterways like this much safer. Computer models designed by the EASAMS company in England, are used to analyze traffic flow in congested waterways. One system was designed for the Lighthouse Authority. It analyzed the number of ships using coastal navigation aids, and predicted what extra aids would be needed in the future.

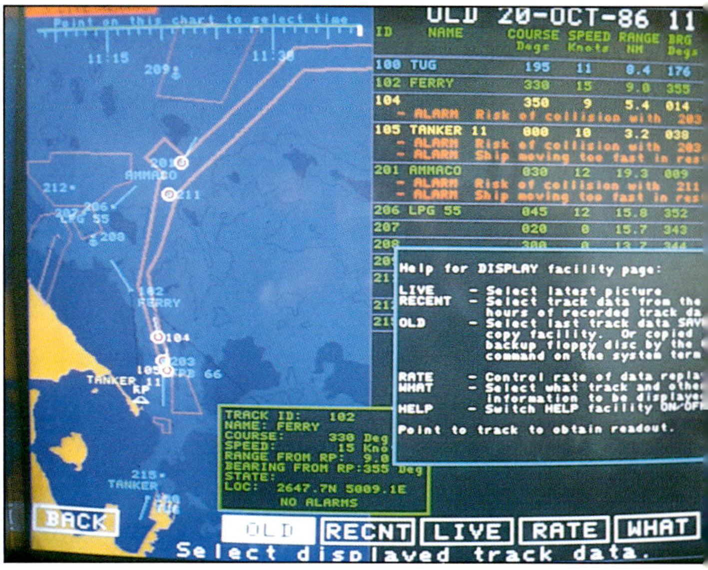

△ The EAMACS harbor surveillance system not only tracks all shipping in the area, but can also plot the rail and road traffic in the harbor area. It is a control system designed for safety, speed and efficiency.

MONITORING THE DANGERS

Measuring pollution levels can be quite difficult. Some have an immediate effect on marine animals. Others accumulate in their bodies for years without seeming to do much obvious harm.

Many different animals are used as indicators of pollution. Limpets are used in studies of oil pollution and the problems caused by some detergents. Oysters reveal the harm being done by some anti-fouling paints used on pleasure boats. The most recent discovery is a tiny shrimp-like creature called a harpacticoid, which can tolerate high levels of most common pollutants but will not live in areas polluted by heavy metals such as mercury and lead.

◁ The diver is injecting oil into the plastic bag to test its effect on the enclosed animal and plant samples.

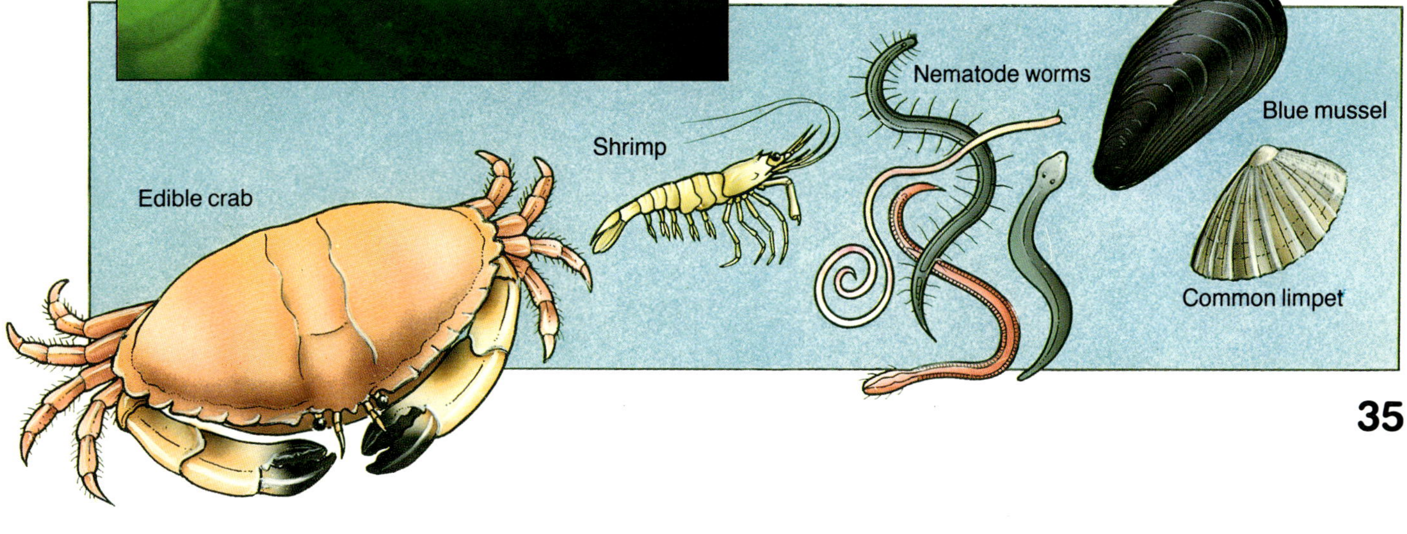

CONTROLLING THE LAND AND SEA

For some countries, golden beaches, blue lagoons and warm seas are a national treasure. Visitors pour in, spending the money that supports the local economy. In other parts of the world, the coast may be exploited in other ways. In parts of Asia, mangroves are cut down for their timber. The desert beaches of Namibia have been torn up in the search for diamonds.

Using the coast in these ways can be dangerous. Upsetting the seashore can increase the damage done by storm waves, and can also ruin the breeding grounds of local fish and birds. There are places, however, where people have no choice. They have to hold back the sea, or even turn the sea into land. In Japan, land is so scarce that the Gobo Power Station was built on an artificial island. In Holland, two-fifths of the land would be flooded if it were not for the massive dikes holding back the sea.

SEA DEFENSES

Ocean waves contain a staggering amount of energy. Every day, they pound the world's shores with as much energy as a 50 megaton bomb, and earlier this century a single storm wave in the English Channel threw a 65 ton concrete block more than 20m (65ft) up the beach at Cherbourg.

It is hardly surprising that cliffs collapse, beach houses are flattened, and even the strongest sea walls sometimes give way in the face of such force. Many different sea wall designs have been tried. Vertical walls are the simplest to build. Curved sea walls hurl the waves up and back on themselves. They protect sea front buildings but can increase beach erosion so that the wall is undermined and eventually collapses. Most of the new designs aim to absorb the waves' energy and so slow them down. They include the dynamic breakwater concept designed in the United States, and the energy-absorbing honeycomb surface designed in England.

ARTIFICIAL ISLANDS

Oil exploration engineers working on the fringes of the Arctic Ocean have been forced to develop many new solutions to drilling problems. The constant motion of sea ice many yards thick would soon demolish a drilling rig standing on the sea bed, and would push a floating rig off position.

△ The "honeycomb" sea wall at Felixstowe in England will absorb 50 percent of the waves' energy. It should last for at least 50 years.

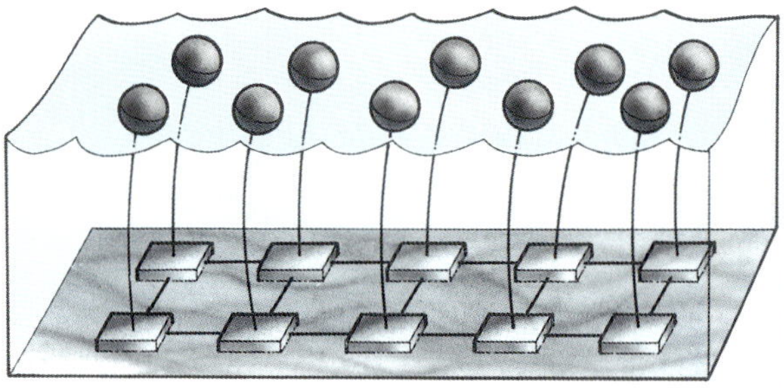

△ The dynamic breakwater consists of floats tethered just below the surface. It can absorb far more energy than a rigid wall.

In spring 1986, Shell drilled their second well in the Sandpiper field in Alaska. The drilling rig stood on an artificial island of gravel, dredged from the surrounding seabed. For a new well at Angasak in the Beaufort Sea, Esso Canada is drilling from an ice island. Water is being pumped on to the ice, where it freezes. The ice gets thicker and thicker until it eventually becomes firmly grounded on the seabed 5.5 m (18 ft) down.

△ The two artificial islands that now form part of the 9 km (5½ mi) barrier were built to provide construction sites for the 65 massive concrete piers.

△ Esso's CRI (Caisson-Retained Island) uses an eight-sided steel wall to hold in the filling of sand, dredged from the surrounding sea bed.

◁ A converted tanker, complete with drill tower, is towed over a sunken steel mat and the mat is partially refloated to join them securely.

HOLDING BACK THE SEA

In October 1986, the last link in Holland's sea defenses was completed. The Ooster Schelde barrier seals off the last opening to the North Sea, and should finally end the threat of flooding that has been with this low-lying country for centuries. More than 40 percent of the country's land has been won back from the sea and lies below sea level. Today, it is protected by 1,300 km (800 mi) of dikes and sea walls.

A solid dam across the Ooster Schelde would have upset the ecology of this huge delta area, so the barrier is designed to remain open for most of the time. When flooding threatens, its 62 steel gates are raised by hydraulic rams.

DELTAPORT

This futuristic design is the work of the "Deltaport" group of St. John's, Newfoundland. It is a complete floating supply and support base and would be anchored near the center of a large oil or gas field. Its hollow shape provides a sheltered port for supply boats, and along one side it has a heliport and an airstrip for STOL (Short Take Off and Landing) aircraft. The steel space-frame structure is filled with buoyancy cylinders.

▽ Scale models of the Deltaport are currently undergoing wave-tank trials.

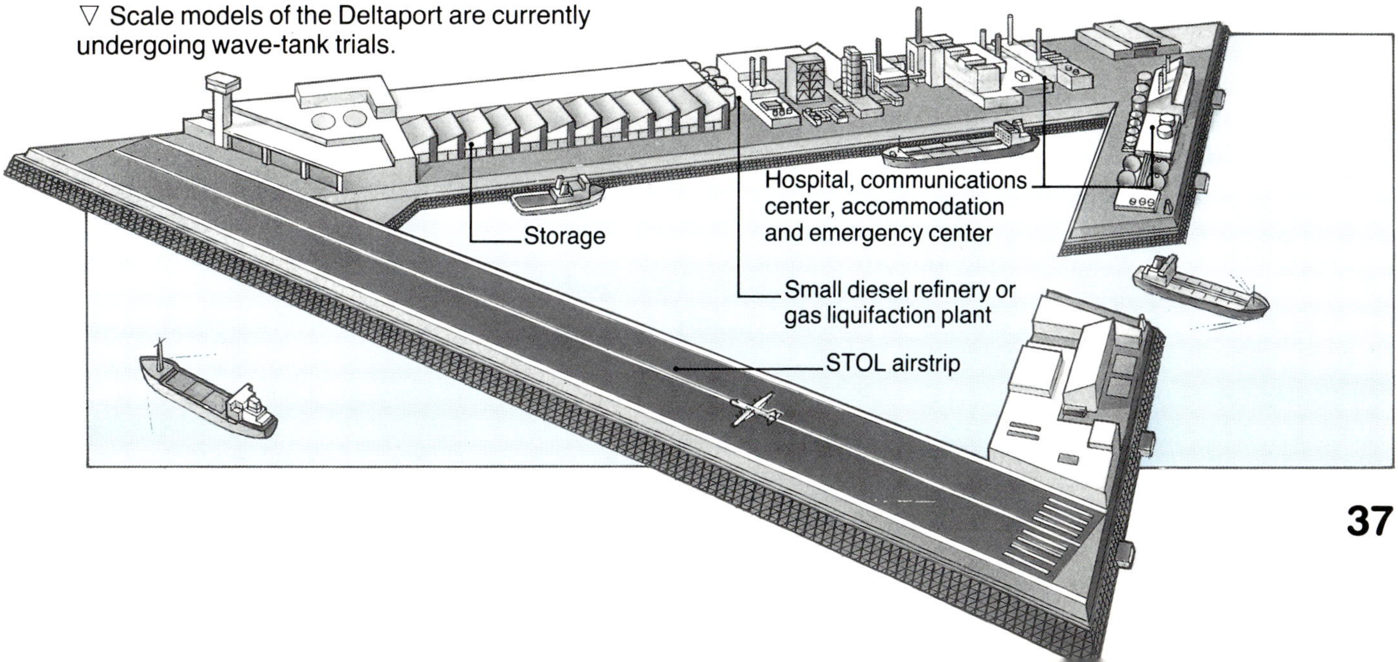

PROTECTING THE MARINE WORLD

The technology that now allows us to exploit the oceans for our own benefit can also help us protect the marine world from harm. In the past, scientists' warnings about environmental damage were often ignored by governments and industries. It was difficult for the scientists to produce enough evidence to prove their case.

Things are different today. There are marine research centers all over the world. Satellites, remote-operated instruments, computerized data processing systems and new laboratory techniques are helping scientists produce detailed, accurate reports on the state of the marine environment. And that can only be good for conservation.

GREENPEACE

The international conservation group Greenpeace uses direct action and hard-hitting publicity campaigns to get its message across. Among its main targets in recent years have been the continued killing of whales, the testing of nuclear weapons on islands in the Pacific, the pollution of Britain's coastline by dumping of untreated sewage and nuclear waste, and plans to farm green turtles in the Indian Ocean.

▽ A typical Greenpeace action – defying cold water jets to get in so close that the ship cannot risk dropping its drums of waste overboard.

WILDLIFE RESEARCH

For many years, scientists and fishermen have argued about whether or not gray seals affect the fish stocks of the North Sea. In 1977, the Department of Agriculture and Fisheries for Scotland claimed that seals were taking 120,000 tons of cod, salmon, plaice and other fish each year. As a result, permits were once again issued for a proportion of the seal population to be culled. However, many biologists believe that the seals feed mainly on sand eels, not the high value food fish. Biologists from the Sea Mammal Research Unit are now experimenting with small radio transmitters that can be glued to a seal's head or back. In this way, the animal can be tracked by the French-American *Argos* satellite system. If the biologists can discover where the seals go, it may prove whether they are guilty or innocent of these charges.

△ Biologists of the British Antarctic Survey are using dummy nests to study feeding and growth patterns in albatross chicks. The glassfiber nest contains an automatic weighing device which monitors the chicks' weight.

THE ENDANGERED GIANTS

The coming of steam-powered catcher boats and explosive harpoons in the 1860s almost meant the end of the whale. At the height of the whaling industry, species such as the Greenland right whale and the bowhead were almost wiped out.

Many countries have now ceased whaling, and there are international quotas for how many whales can be caught by the others. Some species are recovering slowly. Unfortunately, quotas are often ignored and whales are even killed in the interests of "scientific research." Sadly, the International Whaling Commission has no real power, so until *all* nations agree to stop the hunting completely, whales will remain under constant threat of extinction.

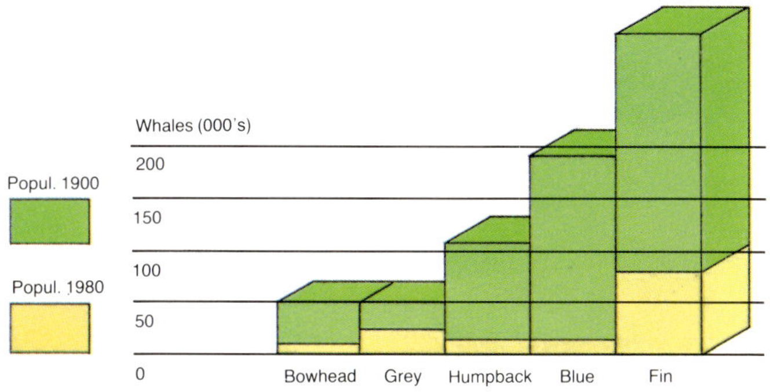

△ Estimated size of whale populations before whaling commenced and at the present time.

△ While the gunner reloads with a barbed harpoon, his assistant inflates this sperm whale's body with an air pipe to keep it afloat.

△ A glass-bottom boat provides visitors to Egypt with a glimpse of the Red Sea's beautiful reefs and their wildlife.

MARINE PARKS AND SANCTUARIES

The idea of submarine national parks is relatively new. The world's first national park was Yellowstone founded in 1872, but it was not until 1960 that the first marine park was set up, in Florida. In 1975, the Australian Government declared the Great Barrier Reef a Marine Park. In 1977, New Zealand established its first Marine Reserve; now it has four. Britain's first Marine Nature Reserve, and so far her only one, was established in 1986. It is the tiny rocky island of Lundy, in the Bristol Channel.

Lundy Island's few square miles of rocky coast may seem unimportant compared with the 230,000 sq km (150,000 sq mi) of coral reefs, islands and lagoons that make up the Great Barrier Reef. But both have their place in preserving the enormous variety of animals and plants that live in the sea. For all its small size, Lundy has a great variety of habitats. It is also home to such rare sea animals as jewel anemones and sunset star corals.

Marine reserves may vary from protected "no-go" areas for people, to open parks in which controlled tourism helps to pay for the park's maintenance.

A NEW DEAL FOR THE OCEANS

In 1988, the world's largest floating hotel was towed into position over Australia's Great Barrier Reef. In addition to hotel facilities, it will offer diving and underwater photography, sport fishing, and tours through the reef in miniature submarines. It is a glimpse into the future. Similar hotels may soon be seen in the Red Sea, the Caribbean and among the islands of Southeast Asia.

Scientists and engineers are now tackling the question of what is to be done with offshore oil platforms when production ceases. Some will have to be removed completely. Others may be broken up on the sea bed and left to provide safe spawning grounds for fish and shellfish. The large concrete platforms might be converted into helipads or used for oceanographic research.

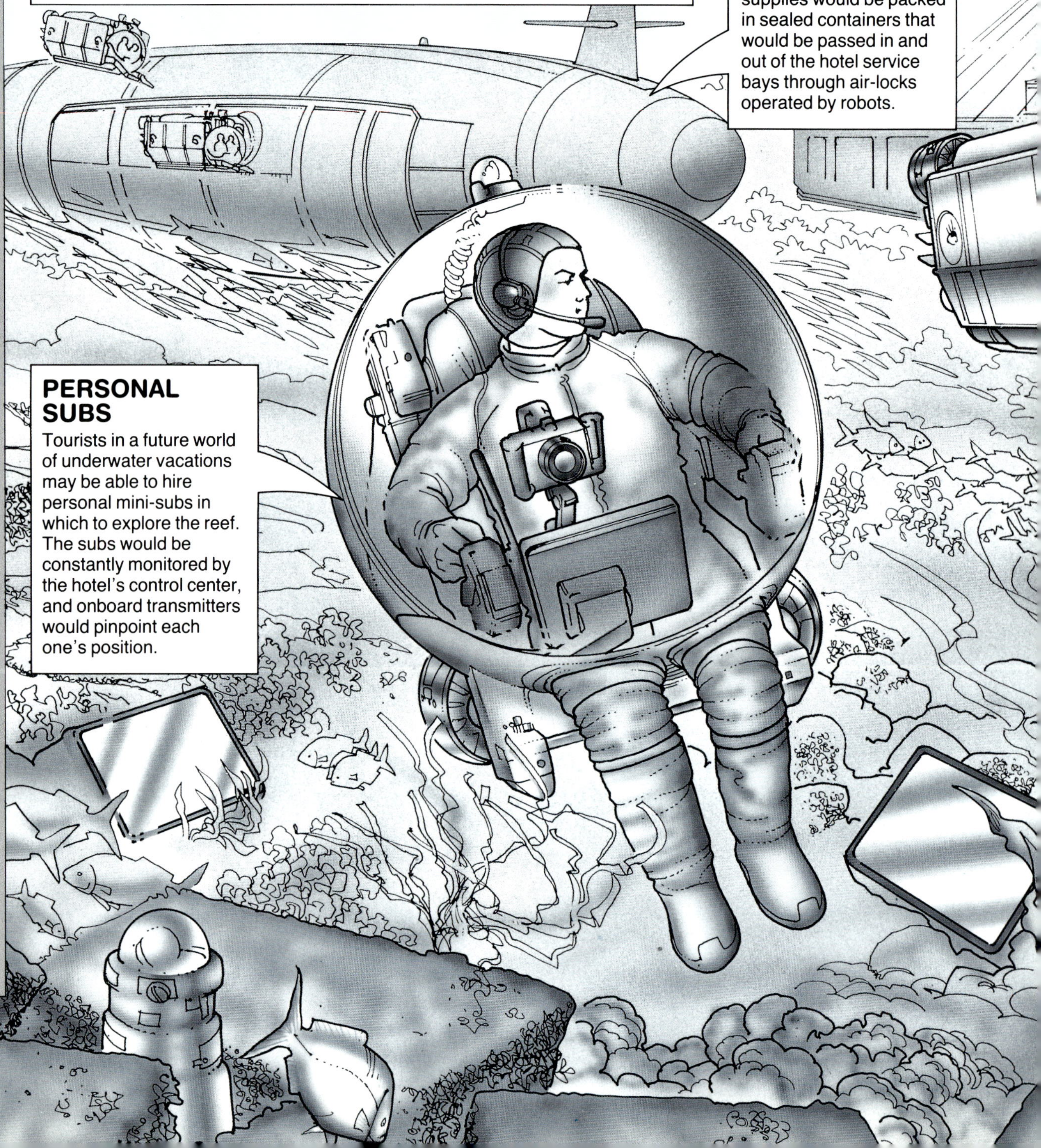

SUPPLY SHUTTLE

Food, clean laundry and other supplies could be brought to a seabed hotel each day by shuttle subs. For quick operation, the supplies would be packed in sealed containers that would be passed in and out of the hotel service bays through air-locks operated by robots.

PERSONAL SUBS

Tourists in a future world of underwater vacations may be able to hire personal mini-subs in which to explore the reef. The subs would be constantly monitored by the hotel's control center, and onboard transmitters would pinpoint each one's position.

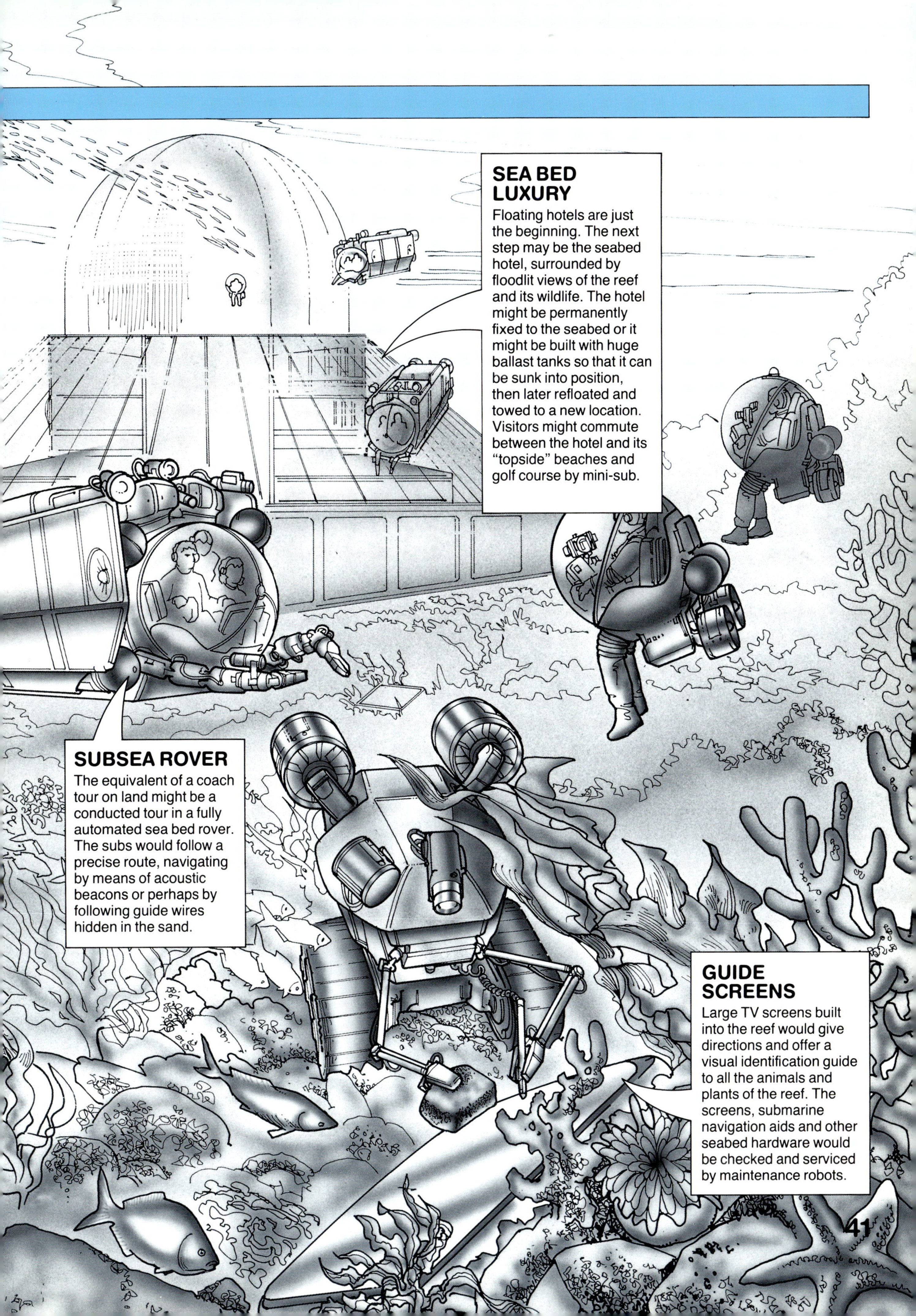

SEA BED LUXURY

Floating hotels are just the beginning. The next step may be the seabed hotel, surrounded by floodlit views of the reef and its wildlife. The hotel might be permanently fixed to the seabed or it might be built with huge ballast tanks so that it can be sunk into position, then later refloated and towed to a new location. Visitors might commute between the hotel and its "topside" beaches and golf course by mini-sub.

SUBSEA ROVER

The equivalent of a coach tour on land might be a conducted tour in a fully automated sea bed rover. The subs would follow a precise route, navigating by means of acoustic beacons or perhaps by following guide wires hidden in the sand.

GUIDE SCREENS

Large TV screens built into the reef would give directions and offer a visual identification guide to all the animals and plants of the reef. The screens, submarine navigation aids and other seabed hardware would be checked and serviced by maintenance robots.

WHAT NEXT?

NEW FRONTIERS

Looking beyond the year 2001, the biggest of all challenges facing politicians, scientists and engineers will be the wise use of the oceans.

Exploiting deep sea minerals may still be several decades away, but already the seas offer a real chance of reducing the amount of hunger and starvation in the world. Coral reefs, for example, are the most productive of all marine habitats. Used carefully, they could produce up to 9 million tons of food every year. New fish farming methods could help feed hungry coastal regions, while biotechnology and new food processing methods may soon provide sources of protein that can be stored for long periods and transported to famine areas without waste.

Perhaps the biggest challenge of all is that the seas can only be used well, and protected from pollution, if all nations cooperate. The first United Nations Conference on the Law of the Sea (UNCLOS I) was held in 1958. Thirty years and many conferences later, the world's nations are still arguing – over fishing zones, mining rights, dumping regulations, and so on. It is slow, but vital, progress.

△ The 140 m (400 ft) long ship is designed to cut through ice a yard thick. It carries a helicopter, and tractors and cranes for unloading onto ice.

ANTARCTIC RESEARCH

In 1988, the world's newest Antarctic research ship goes into service, supplying the U.S.S.R.'s polar research bases. The new ship is equipped for oceanographic research. Echo sounders will map the sea bed in depths down to 10 km (6 mi). High-altitude weather balloons will be tracked up to 25 km (15 mi) into the atmosphere. The ship's 10 onboard laboratories will carry out research into sea chemistry, marine life, tide and current movements and wave motion.

NEW RESEARCH TECHNIQUES

Scientists at the Scripps Institution, the Woods Hole Oceanographic Institution and the Massachussetts Institute of Technology have built an acoustic system for tracking temperature variations and currents. The 1600 kg (3,500 lb) device is tethered 750 m (2,500 ft) below the surface and transmits a humming sound that can be picked up, about 11 minutes later, by another instrument 1,000 km (620 mi) away. It will track the swirling eddies of warm and cold water that are the ocean's weather systems.

Other developments include IOS's VAESAT buoy, which collects surface current information and transmits it to Britain and France via the *Argos* satellite; TOBI, a new deep-towed sonar scanner; and FIDO, an instrument that samples the tiny solid particles in seawater.

Commercial uses for new technology include computerized traffic control systems and dyes and biological tracers for checking the cleanliness of coastal waters.

◁ A camera and light assembly for monitoring the activity of animals contained in the mesh cage.

▷ Young European flatfish of the turbot family being fed in a tank on one of the experimental fish farms run by the BP company.

▲ FOOD FOR THE FUTURE

Cities beneath the sea will probably remain a science fiction dream, but factories beneath the sea may be a real possibility. The rapid growth of biotechnology, using bacteria to perform complex biological processes, could result in new ways of producing fuel gas, or fertilizers and medical compounds. And if these essential products can be made from raw materials from the sea, why not place the factory unit right there on the seabed?

In the meantime, research goes on into ways of producing more food-fish from the sea. Australian scientists are investigating the best materials for building artificial reefs for shellfish farming. Japanese biologists are exploring ways of using cold, nutrient-rich, unpolluted water from the deep offshore zone to make Japan's shallow coastal waters and fishfarms more productive.

▼ CONSERVING THE RESOURCE

Scientists have described coral reefs as the most complex and fascinating of all marine habitats. They are very ancient. There were coral reefs 450 million years ago, and individual reefs may last for thousands of years. They are built by living animals, yet they are as hard as rock—a kind of living geology. Reefs are the richest habitats in the sea, the marine equivalent of a tropical forest, with up to 3,000 different species in the space of a few miles. All this makes them a priceless resource. They are a rich source of food, and they are a huge tourist attraction.

◁ Reefs can only be protected if they are correctly understood. Here, in the Red Sea, a biologist is studying the life processes of living corals. By enclosing the samples beneath a Perspex dome he can monitor the temperature, salinity and gas content of the surrounding water.

KENMORE EAST H S LIBRARY

GLOSSARY

Acoustic Related to sound or to the sense of hearing. An acoustic signal is a pulse of sound transmitted through the water.

Alginates A group of chemical substances that occur naturally in seaweeds.

Asdic An early form of sonar equipment. The name is from the **A**nti-**S**ubmarine **D**etection **I**nvestigation **C**ommittee.

Beacon A signal station. The first beacons were fires on top of hills. Modern versions use radio, radar or acoustic signals to assist navigation.

Biotechnology The new technology of using living things, usually bacteria, to carry out chemical processes that scientists cannot do, or can only do with huge, complicated apparatus.

Compression (and **decompression**) The process of slowly increasing the pressure on a diver's body so that it will match the pressure at the depth where he or she is going to work.

Crust The outermost layer of the Earth. The crust is a thin, rigid, stony layer, which floats on a semi-liquid layer called the mantle.

Cull To kill a certain number of animals in a population in order to control the population and prevent it from becoming too big for the available habitat or supply.

Environment Everything that makes up our surroundings – the earth, air, water and wildlife.

Fish meal Dried, ground-up fish, widely used as an animal feed and fertilizer.

Food technology Anything to do with the production, processing, handling and packaging of food for human use.

Genetic engineering Altering the features of an animal or plant by deliberately changing its DNA ("genetic blueprint") code.

Habitat The place where an animal or plant normally lives.

Hydrofoil A ship that can increase its speed by raising its hull out of the water on underwater wings so that drag is reduced.

Kelp A family of large, brown seaweeds. The biggest grow to 30 m (100 ft) or more in length.

Krill A small shrimp-like animal that lives in the cold southern oceans and is the main food of many whale and fish species.

Laser A device that produces a beam of very pure, intense light. Pulses of laser light can be sent along optical fibres to transmit TV pictures and computer information.

Monitor *noun*: A computer screen or Visual Display Unit. *verb*: To keep a constant or regular check on something.

Pollution Anything that spoils or contaminates the natural environment. The atmosphere is polluted by smoke and industrial waste gases; rivers may be polluted by sewage, rubbish, chemical spills and so on.

Radar **R**adio **D**irection **A**nd **R**anging. A system invented in World War II for locating aircraft by transmitting radio waves and picking up the echoes that bounce back from them.

Remote control Control of a vehicle or any other machine from some distance away, either by radio signals, or acoustic signals, or by sending instructions electronically via a connecting wire or cable.

Resource A natural stock or supply of something that can be used by people. Energy resources, for example, include coal, wave and thermal energy.

Robot A machine that can be programmed to carry out certain movements and actions automatically.

Sonar **S**ound **N**avigation and **R**anging. A system very similar to radar except that it uses transmitted sound waves and their echoes instead of radio waves.

Spawn *verb*: To produce eggs. The word is used when talking about fish, frogs and shellfish. *noun*: Fish, frog or shellfish eggs.

STOL **S**hort **T**ake-**O**ff and **L**anding. A type of aircraft designed to operate from very short runways.

Submersible A vessel designed to operate under water for short periods. Used for scientific research, underwater repairs and construction work, and submarine rescue.

Surimi A clear colorless gel of protein made by repeatedly washing and pressing fish meat.

INDEX

acoustic signals 10, 11, 43
alginates 22
altimeter, radar 10
Alvin 12, 14, 15, 29
Antarctic research 42
aqualungs 14
archaeology, underwater 14–15
Argo 15
artificial islands 36
Atlantis II 12

Bathysnap 11
biology, marine 12, 20, 22, 31, 35, 39
breathing underwater 16
buoyancy floats 11

Challenger, HMS 8
chemosynthesis 31
Coastal Zone Color Scanner (CZCS) 18
command cable 13, 14
concretions 14
conservation 20, 21, 38, 39
coral reefs 43

deep ocean basins 12
Deep Rover 16
Deep Submergence Research Vehicle (DSRV) 14
Defense, The 14
Deltaport 37
derrick 11
desalination 28
divers 12, 13, 17, 43
drilling 10
  deep-sea 11
  for oil 6
  rigs 12, 13, 26, 36

electronic beacons 16
ERS-1 satellite 10

factory ships 17, 19
fish farming 21, 30
fish processing 23, 25
fishery patrols 20
fishing methods 18, 19, 20
fossils 11

genetic engineering 31
Geological Long Range Inclined Asdic (GLORIA) 10
geology 10, 12
giant tube worms 31
Great Barrier Reef 39
Greenpeace 38

horseshoe crab 30
hotels, underwater 41
hot-water vents 12

icebergs 28
Institute of Oceanographic Studies (IOS) 10

Japan Marine Science and Technology Center (JAMSTEC) 12
Jason Jr 14, 15
JOIDES Resolution 11

Kaiyo 12
kelp 22, 24
krill 23

marine parks and reserves 39
Mary Rose 14
mineral nodules 29
mini-subs 13, 40
motors, water-driven 12
mud waves 10

National Oceanographic and Atmospheric Administration (NOAA) 18
Nereis poison 30
nets, fishing 10, 17, 18, 21, 23

ocean floor 10, 11
Ocean Thermal Energy Conversion (OTEC) 27
oil spills 34
oilfields 26
Oscillating Water Column 27
over-fishing 20

PEG wood treatment 15
pipelines, undersea 8, 12, 13
plankton 18
pollution 35
pop-up technology 11
positioning systems 11, 12

Remote Maintenance Vehicle (RMV) 26
Remote Operated Vehicles (ROVs) 13, 15, 16
research vessels 12, 42
rock studies 11

salinity 8, 28
satellites 10, 18
Scripps Institution of Oceanography 20, 29, 43
sea defences 36, 37
seabed 10, 29
  animals 11, 31, 45
  mapping 10
  smokers 29
seafood 18–19, 21, 22–25, 45
SEASAT 10
seaweed 22, 24, 25, 31
shipbuilders 12
Small Waterplane Area Twin Hull (SWATH) vessels 12
smokers 29
sonar 14, 16, 17
squid jiggers 19
submarine volcanoes 10
submarines 12, 41
submersibles 8, 12, 13, 17, 29, 40
surimi 23, 25
Synthetic Natural Gas (SNG) 24

Titanic, the 14, 15
tourists 13, 39
traffic control, sea 35, 43

underwater
  bases 16, 40–41
  communications 12
  exploration 10–11
United Nations Conference on the Law of the Sea (UNCLOS I) 42

waste dumping 34
wave power 27, 32
whales 39
wildlife research 38
wind power 32
Woods Hole Oceanographic Institution 14
wrecks 15

32147

551.46 Bramwell, Martyn
B
Oceanography

$19.14

DATE			

BAKER & TAYLOR BOOKS